Cells, Cells and Nothing but Cells: Discoveries, Challenges and Directions

Cells, Cells and Nothing but Cells: Discoveries, Challenges and Directions

Editor

Alexander E. Kalyuzhny

MDPI • Basel • Beijing • Wuhan • Barcelona • Belgrade • Manchester • Tokyo • Cluj • Tianjin

MDPI

Editor
Alexander E. Kalyuzhny
University of Minnesota
Twin Cities
USA

Editorial Office
MDPI
St. Alban-Anlage 66
4052 Basel, Switzerland

This is a reprint of articles from the Proceedings published online in the open access journal *Biology and Life Sciences Forum* (ISSN 2673-9976) (available at: https://www.mdpi.com/2673-9976/21/1).

For citation purposes, cite each article independently as indicated on the article page online and as indicated below:

LastName, A.A.; LastName, B.B.; LastName, C.C. Article Title. *Journal Name* **Year**, *Volume Number*, Page Range.

ISBN 978-3-0365-7852-1 (Hbk)
ISBN 978-3-0365-7853-8 (PDF)

Contents

biology and life sciences forum

Abstract

New Approaches Targeting the Invasive Phenotype of Prostate Cancer-Associated Fibroblasts †

Marzia Di Donato *, Pia Giovannelli, Antimo Migliaccio and Gabriella Castoria

Department of Precision Medicine, University of Campania "Luigi Vanvitelli", 80138 Naples, Italy; pia.giovannelli@unicampania.it (P.G.); antimo.migliaccio@unicampania.it (A.M.); gabriella.castoria@unicampania.it (G.C.)

* Correspondence: marzia.didonato@unicampania.it

† Presented at Cells, Cells and Nothing but Cells: Discoveries, Challenges and Directions, 6–8 March 2023; Available online: https://sciforum.net/event/cells2023.

Abstract: Prostate cancer (PC) is one of the most widespread malignancies among males worldwide. The androgen receptor (AR) drives its development and progression and still represents the main target of PC therapy. Second-generation antiandrogens have, indeed, improved the patient's management. Nonetheless, hormone resistance and tumour progression frequently develop. While the majority of drugs currently used in PC target the AR functions in epithelial PC cells, the role of the receptor in PC-associated fibroblasts (CAFs) and PC progression remains unresolved, and only a few therapeutics affecting the stromal AR functions have been developed so far. By combining several approaches, we have shown that AR associates with Filamin A (FLNa), thus promoting migration and invasion of androgen-challenged CAFs from PC patient's specimens at different Gleason's scores. By using 2D and 3D cultures, we have demonstrated that CAFs move towards epithelial PC cells and promote the increase in PC organoid size. The stapled peptide Rh-2025u disrupts the androgen-triggered AR/FLNa complex assembly and impairs these responses in monolayer cells as well as 3D models. Furthermore, it reduces the overall tumour area in androgen-treated 3D co-culture. Mechanistically, our findings posit that AR/FLNa complex recruits β1 integrin and the membrane type-matrix metalloproteinase 1 upon the androgen challenging of CAFs. The activation of a protease cascade leading to extracellular matrix (ECM) remodelling then follows. Rh-2025u peptide interferes in the assembly of this multimolecular complex and impairs ECM remodelling. As such, CAFs can no longer navigate through ECM. In summary, we propose the Rh-2025u peptide as a new drug, which alone or in combination with other emerging therapies may allow a more rational treatment of PC. Pharmacological blockade of AR functions in CAFs is indeed neglected and the approach we propose would improve the treatment's outcome in PC patients.

Keywords: prostate cancer; carcinoma-associated fibroblasts; invasion; 3D models

Citation: Di Donato, M.; Giovannelli, P.; Migliaccio, A.; Castoria, G. New Approaches Targeting the Invasive Phenotype of Prostate Cancer-Associated Fibroblasts. *Biol. Life Sci. Forum* **2023**, *21*, 1. https://doi.org/10.3390/blsf2023021001

Academic Editor: Alexander E. Kalyuzhny

Published: 16 March 2023

Author Contributions: Conceptualization, G.C. and M.D.D.; methodology, M.D.D. and P.G.; software, A.M.; investigation, M.D.D. and P.G.; writing—original draft preparation, M.D.D. and G.C.; writing—review and editing, M.D.D. and G.C.; supervision, A.M. and G.C. All authors have read and agreed to the published version of the manuscript.

Funding: This research was funded by Italian Ministry of University and Scientific Research (P.R.I.N. 2017EKMFTN_002 to G.C.), Regione Sicilia (Progetto di Ricerca Finalizzata RF-2019-12368937 to A.M.) and Vanvitelli Young Researcher (PATG.Rice.Base.GiovaniRicercatori2022.IDEA to M.D.D.).

Institutional Review Board Statement: Not applicable.

Informed Consent Statement: Not applicable.

Data Availability Statement: Not applicable.

Conflicts of Interest: The authors declare no conflict of interest.

biology and life sciences forum

MDPI

Abstract

Microgravity Exposure Alterations of Cellular Junctions Proteins in TCam-2 Cells: Localization and Interaction †

Marika Berardini [1], Luisa Gesualdi [1], Francesca Ferranti [2], Maria Addolorata Mariggiò [3], Caterina Morabito [3], Simone Guarnieri [3], Giulia Ricci [4] and Angela Catizone [1,*]

1 Department of Anatomy, Histology, Forensic-Medicine and Orthopedics, Section of Histology and Embryology, "Sapienza" University of Rome, 00161 Rome, Italy; marika.berardini@uniroma1.it (M.B.); luisa.gesualdi@uniroma1.it (L.G.)
2 Human Spaceflight and Scientific Research Unit, Italian Space Agency, 00161 Rome, Italy; francesca.ferranti@asi.it
3 Department of Neuroscience, Imaging and Clinical Sciences-CAST, "G. d'Annunzio" University of Chieti-Pescara, 66100 Chieti, Italy; mariggio@unich.it (M.A.M.); cmorabit@unich.it (C.M.); simone.guarnieri@unich.it (S.G.)
4 Department of Experimental Medicine, Università degli Studi della Campania "Luigi Vanvitelli", 80138 Naples, Italy; giulia.ricci@unicampania.it
* Correspondence: angela.catizone@uniroma1.it

† Presented at Cells, Cells and Nothing but Cells: Discoveries, Challenges and Directions, 6–8 March 2023; Available online: https://sciforum.net/event/cells2023.

Abstract: One of the most important hazards of the space environment is microgravity, which causes an alteration in the physiology of different systems, including the reproductive one. It is widely accepted that cytoskeleton is the microgravity-sensitive apparatus of the cells, and that cytoskeletal modifications are responsible for microgravity-triggered cell alterations. We established a 3D free-floating culture system from TCam-2 cell, a human seminoma cell line, and then exposed the obtained TCam-2 spheroids for 24 h at unitary gravity (UG), or under a simulated microgravity condition (SM), using the random position machine (RPM). We tested the cytoskeletal and junctional features of these samples using Western blot and confocal microscopy analysis to elucidate the impact of microgravity on the adherent and occluding junctions of TCam-2 spheroids. The junctional ultrastructure was studied using transmission electron microscopy (TEM). TEM analysis revealed the presence of occluding junctions both in UG or SM samples. Even if Western blot revealed no quantitative difference in actin and occludin proteins both in UG and SM exposed samples, fluorescence colocalization analysis showed a significative increase in the colocalization area of occludin and actin proteins in the superficial layer of TCam-2 spheroids grown in RPM conditions. This result let us speculate that tight junction functionality is different in UG and SM exposed spheroids. As far as adherent junctions are concerned, TEM analysis revealed adherent junctions both in UG or SM samples. Moreover, we observed by Western blot a trend in terms of the increase in the vimentin expression in SM exposed spheroids. Confocal microscopy analyses confirmed this significant increase. All together, these data suggest that simulated microgravity conditions in TCam-2 spheroids alter the tight junction assembly, while the increase in the intermediate filament's structures can in part be associated with an enrichment in the adherent junctions. A functional investigation is needed to more deeply clarify this hypothesis.

Keywords: microgravity; cytoskeleton; TCam-2 cell

Citation: Berardini, M.; Gesualdi, L.; Ferranti, F.; Mariggiò, M.A.; Morabito, C.; Guarnieri, S.; Ricci, G.; Catizone, A. Microgravity Exposure Alterations of Cellular Junctions Proteins in TCam-2 Cells: Localization and Interaction. *Biol. Life Sci. Forum* **2023**, *21*, 2. https://doi.org/10.3390/blsf2023021002

Academic Editor: Alexander E. Kalyuzhny

Published: 16 March 2023

Author Contributions: Conceptualization, A.C., G.R., L.G. and M.B.; methodology, M.B., L.G., M.A.M., C.M. and S.G.; software, A.C., G.R., M.B. and L.G.; validation, A.C., G.R., M.B. and L.G.; formal analysis, A.C., G.R., M.B. and L.G.; investigation, A.C., G.R., M.B. and L.G.; resources, A.C., G.R., M.B. and L.G.; data curation, A.C., G.R., M.B. and L.G.; writing—original draft preparation,

A.C. and G.R.; writing—review and editing, M.B., L.G., M.A.M., C.M. and S.G.; visualization, A.C. and G.R.; supervision, F.F.; project administration, A.C. and F.F.; funding acquisition, A.C. and G.R. All authors have read and agreed to the published version of the manuscript.

Funding: This research was funded by grants from ASI–Italian Space Agency, number 2020-24-HH.0.

Institutional Review Board Statement: Not applicable.

Informed Consent Statement: Not applicable.

Data Availability Statement: Not applicable.

Conflicts of Interest: The authors declare no conflict of interest.

biology and life sciences forum

Abstract

Maternal High Fat Diet Multigenerationally Impairs Hippocampal Adult Neurogenesis †

Francesca Natale [1,2], Matteo Spinelli [1,2], Saviana Antonella Barbati [1], Lucia Leone [1,2], Salvatore Fusco [1,2,*] and Claudio Grassi [1,2]

1 Department of Neuroscience, Università Cattolica del Sacro Cuore, 00168 Rome, Italy; francesca.natale1@unicatt.it (F.N.); matteo.spinelli@unicatt.it (M.S.); savianaantonella.barbati@unicatt.it (S.A.B.); lucia.leone@unicatt.it (L.L.); claudio.grassi@unicatt.it (C.G.)
2 Fondazione Policlinico Universitario A. Gemelli IRCCS, 00168 Rome, Italy
* Correspondence: salvatore.fusco@unicatt.it
† Presented at Cells, Cells and Nothing but Cells: Discoveries, Challenges and Directions, 6–8 March 2023; Available online: https://sciforum.net/event/cells2023.

Abstract: Metabolic dysregulation harms brain health. Early-life (pre- and perinatal) dysmetabolic stimuli have been demonstrated to affect central nervous system (CNS), multigenerationally impairing brain plasticity and cognitive functions in adult offsprings. In our previous work, we reported that maternal high fat diet (HFD) impaired synaptic plasticity, learning and memory of descendants until the third generation. Neural stem and progenitor cells (NSPCs) represent the cellular source of newborn neurons in the subgranular zone of the hippocampus, and their fate is finely modulated by metabolic signals. Epigenetic mechanisms are key factors controlling the neural fate of NSPCs and they dynamically regulate CNS development and adult neurogenesis. Here, we demonstrate that progenitor HFD altered both the proliferation of NSPCs and the hippocampal adult neurogenesis on second and third generations of progeny (F2HFD and F3HFD), leading to the depletion of neurogenic niche in the descendants. Moreover, in NSPCs isolated from the hippocampus of HFD descendants we found reduced expression of genes regulating stem cell proliferation and neuro-differentiation (i.e., Hes1, NeuroD1, Bdnf). Furthermore, maternal HFD-related metabolic stress induced a rearrangement of STAT3/5 transcription factors occurring on the regulatory sequences of NeuroD1 and Gfap genes, causing the epigenetic repression of pro-neurogenic and the activation of pro-glial differentiation genes. Collectively, our data indicate that maternal HFD multigenerationally impairs hippocampal neural stem cell niche via epigenetic inhibition of pro-neurogenic gene expression in NSPCs.

Keywords: hippocampal adult neurogenesis; neural stem and progenitor cells; epigenetics; maternal HFD

Citation: Natale, F.; Spinelli, M.; Barbati, S.A.; Leone, L.; Fusco, S.; Grassi, C. Maternal High Fat Diet Multigenerationally Impairs Hippocampal Adult Neurogenesis. *Biol. Life Sci. Forum* **2023**, *21*, 3. https://doi.org/10.3390/blsf2023021003

Academic Editor: Alexander E. Kalyuzhny

Published: 17 March 2023

Author Contributions: Conceptualization, S.F. and C.G.; formal analysis, S.F.; funding acquisition, S.F. and C.G.; investigation, F.N., M.S. and S.A.B.; methodology, L.L. and S.F.; project administration, S.F.; supervision, S.F. and C.G.; validation, F.N. and M.S.; visualization, F.N.; writing—original draft, F.N. and S.F.; writing—review and editing, S.F. and C.G. All authors have read and agreed to the published version of the manuscript.

Funding: This research was funded by the Italian Ministry of University and Research (SIR 2014 RBSI14ZV59 to S.F.) and by Ministero della Salute—Ricerca Corrente 2022 Fondazione Policlinico Universitario A. Gemelli IRCCS (to C.G.).

Institutional Review Board Statement: The animal study protocol was approved by the Ethics Committee of Università Cattolica del Sacro Cuore (authorization n. 39/2017-PR of 16 January 2017) and were fully compliant with Italian (Ministry of Health guidelines, Legislative Decree No. 116/1992) and European Union (Directive No. 86/609/EEC) legislations on animal research. The methods were carried out in strict accordance with the approved guidelines.

Informed Consent Statement: Not applicable.

Data Availability Statement: The data presented in this study are available as Ref. [1].

Conflicts of Interest: The authors declare no conflict of interest.

Reference

1. Natale, F.; Spinelli, M.; Barbati, S.A.; Leone, L.; Fusco, S.; Grassi, C. High Fat Diet Multigenerationally Affects Hippocampal Neural Stem Cell Proliferation via Epigenetic Mechanisms. *Cells* **2022**, *11*, 2661. [CrossRef] [PubMed]

biology and life sciences forum

MDPI

Abstract

The Role of the Androgen Receptor in Skeletal Muscle and Its Utility as a Target for Restoring Muscle Functions †

Carmela Sorrentino [1,‡], Giulia Gentile [1,‡], Rosa D'Angiolo [1], Carmela Barra [1], Ferdinando De Stefano [1], Fabrizio Licitra [1], Emilia Sabbatino [1], Viviana Tutino [1], Antimo Moretti [2], Pia Giovannelli [1], Giovanni Iolascon [2], Antimo Migliaccio [1], Gabriella Castoria [1] and Marzia Di Donato [1,*]

1 Department of Precision Medicine, University of Campania "L. Vanvitelli", 80138 Naples, Italy; carmela.sorrentino@unicampania.it (C.S.); giuliagentile93@gmail.com (G.G.); rosa.dangiolo@unicampania.it (R.D.); carme.barra@studenti.unina.it (C.B.); ferdinando.destefano2@studenti.unicampania.it (F.D.S.); fabrizio.licitra@unicampania.it (F.L.); emiliasabbatino@gmail.com (E.S.); v.tutino@studenti.unina.it (V.T.); pia.giovannelli@unicampania.it (P.G.); antimo.migliaccio@unicampania.it (A.M.); gabriella.castoria@unicampania.it (G.C.)

2 Multidisciplinary Department of Medical-Surgical and Dental Specialties, University of Campania "L. Vanvitelli", 80138 Naples, Italy; antimo.moretti@unicampania.it (A.M.); giovanni.iolascon@unicampania.it (G.I.)

* Correspondence: marzia.didonato@unicampania.it

† Presented at Cells, Cells and Nothing but Cells: Discoveries, Challenges and Directions, 6–8 March 2023; Available online: https://cells2023.sciforum.net/.

‡ These authors contributed equally to this work.

Abstract: Aging is accompanied by a progressive decrease in skeletal muscle mass and function. This process is characterized by the decrease of sex steroid hormone levels due to andropause and menopause. The axis androgens/androgen receptor (AR) sustains muscle size through classic (also called genomic) and non-classic (or non-genomic) actions to elicit various biological responses. Non-genomic androgen effects act through the crosstalk of AR with other partners. Recently, a specific interaction has been shown to occur through AR and filamin A or Src in different types of normal and malignant cells. From these interactions, the activation of several downstream effectors (paxillin, FAK, MAPK, Akt) follows. Such events induce cell proliferation and survival as well as metabolic changes. Irrespective of the sex of the individual, the more important signaling hubs linking the AR non-genomic circuit with cytoskeleton organization have been analyzed by the Western blot of lysate proteins from human skeletal muscle biopsies (obtained from both young and old patients) and C2C12 skeletal muscle cells. The phosphorylation of filamin A and paxillin increases in biopsies derived from old patients (>61 years), as compared with those derived from young patients (<58 years). Furthermore, AR is weakly expressed in samples from old patients, as compared with young patients. Consistent with these findings, C2C12 cells express abundant amounts of AR that increase during the differentiation. This latter finding suggests an involvement of the androgen-triggered rapid activation of several signaling effectors (e.g., MAPK, Akt, Src, FAK) in skeletal muscle disease. Taken together, our findings suggest that the downregulation of the androgen signaling, or of the AR expression, is a key node in the pathogenesis of skeletal muscle related to aging, and is thus related to excessive metabolic functions and loss of skeletal muscle. Given the important knowledge gaps with regards to the mechanism by which androgens regulate skeletal mass functions, more research is needed. Other in vitro or in vivo experiments are necessary in order to inform the utility of targeting the non-genomic AR signaling pathways with new selective androgen receptor modulators and to improve the clinical outcome of age-related frailty and sarcopenia.

Keywords: skeletal muscle; aging; androgen receptor; androgens; cell signaling

Citation: Sorrentino, C.; Gentile, G.; D'Angiolo, R.; Barra, C.; De Stefano, F.; Licitra, F.; Sabbatino, E.; Tutino, V.; Moretti, A.; Giovannelli, P.; et al. The Role of the Androgen Receptor in Skeletal Muscle and Its Utility as a Target for Restoring Muscle Functions. *Biol. Life Sci. Forum* **2023**, *21*, 5. https://doi.org/10.3390/blsf2023021005

Academic Editor: Alexander E. Kalyuzhny

Published: 20 March 2023

Author Contributions: Conceptualization, G.C., M.D.D., G.I. and A.M. (Antimo Moretti); methodology, C.S., G.G., R.D., F.D.S., F.L., E.S., C.B. and V.T.; software, A.M. (Antimo Migliaccio), A.M. (Antimo Moretti) and C.S.; writing—original draft preparation, M.D.D. and G.C.; writing—review and editing, M.D.D., P.G. and G.C. All authors have read and agreed to the published version of the manuscript.

Funding: This research was funded by the Italian Ministry of University and Scientific Research (P.R.I.N. 2017EKMFTN_002 to G.C.), Regione Sicilia (Progetto di Ricerca Finalizzata RF-2019-12368937 to A.M.) and Vanvitelli Young Researcher (PATG.Rice.Base.GiovaniRicercatori2022.IDEA to M.D.D.).

Institutional Review Board Statement: Not applicable.

Informed Consent Statement: Not applicable.

Data Availability Statement: Not applicable.

Conflicts of Interest: The authors declare no conflict of interest.

biology and life sciences forum

MDPI

Abstract

Microgravity Exposure Induces Antioxidant Barrier Deregulation and Mitochondrial Structure Alterations in TCam-2 Cells †

Luisa Gesualdi [1], Marika Berardini [1], Francesca Ferranti [2], Anna Reale [3], Michele Zampieri [3], Katsiaryna Karpach [3], Maria A. Mariggiò [4], Caterina Morabito [4], Simone Guarnieri [4], Angela Catizone [1] and Giulia Ricci [5,*]

1 Department of Anatomy, Histology, Forensic-Medicine and Orthopedics, Section of Histology and Embryology, Sapienza University of Rome, 00161 Roma, Italy; luisa.gesualdi@uniroma1.it (L.G.); marika.berardini@uniroma1.it (M.B.); angela.catizone@uniroma1.it (A.C.)

2 Human Spaceflight and Scientific Research Unit, Italian Space Agency, 00133 Roma, Italy; francesca.ferranti@asi.it

3 Department of Experimental Medicine, "Sapienza" University of Rome, 00161 Roma, Italy; anna.reale@uniroma1.it (A.R.); michele.zampieri@uniroma1.it (M.Z.); karpach.1792414@studenti.uniroma1.it (K.K.)

4 Department of Neuroscience, Imaging and Clinical Sciences-CAST, "G.d'Annunzio" University of Chieti-Pescara, 66100 Chiet, Italy; maria.mariggio@unich.it (M.A.M.); caterina.morabit@unich.it (C.M.); simone.guarnieri@unich.it (S.G.)

5 Department of Experimental Medicine, Università degli Studi della Campania "Luigi Vanvitelli", 80138 Napoli, Italy

* Correspondence: giulia.ricci@unicampania.it

† Presented at Cells, Cells and Nothing but Cells: Discoveries, Challenges and Directions, 6–8 March 2023; Available online: https://cells2023.sciforum.net/.

Abstract: One of the hallmarks of microgravity-induced alterations in several cell models is an alteration in oxidative balance. Notably, male germ cells, sensitive to oxidative stress, have also been shown susceptibility to changes in gravitational force. To gain more insights into the mechanisms of male germ cells' response to altered gravity, a 3D cell culture model was established from TCam-2 cells, a seminoma cell line and the only available in vitro model to study mitotically active human male germ cells. TCam-2 spheroids were cultured for 24 hours under unitary gravity (UG) or simulated microgravity conditions (SM), which was achieved using a random positioning machine (RPM). Apoptosis and necrosis analyses performed on the UG- and SM exposed samples revealed no significant differences in all of the cell death markers. Notably, the Mitosox assay revealed significant oxidation of mitochondria, after microgravity exposure, at least at this culture time. In the SM-treated samples, gene expression levels (evaluated by real-time PCR) of the main enzymes of the antioxidant barrier, GPX1 and NCF1, were reduced, indicating an influence of SM on mitochondrial function. Notably, the expression of HMOX, involved in the heme catabolism of mitochondrial cytochromes, was increased. The SOD, XDH, CYBA, NCF-2, TXN, and TXNRD genes were not affected. The ultrastructural analysis by transmission electron microscopy revealed that SM significantly altered TCam-2 spheroid mitochondria, which appeared swollen and, in some cases, disrupted. Indeed, mitophagy, or mitochondrial autophagy, appears to be more represented in the samples exposed to simulated microgravity. This result seems to be in line with the increase, mediated by the simulated microgravity, in the enzyme HMOX. All together, these preliminary data demonstrate TCam-2 spheroids' sensitivity to acute SM exposure, strongly indicating a microgravity-dependent modulation of mitochondrial morphology and activity and encouraging us to perform further investigations on the chronical exposure to SM of TCam-2 spheroids.

Keywords: mitochondria; simulated microgravity; cellular spheroids; TCam-2 cells; oxidative stress; mitophagy

Citation: Gesualdi, L.; Berardini, M.; Ferranti, F.; Reale, A.; Zampieri, M.; Karpach, K.; Mariggiò, M.A.; Morabito, C.; Guarnieri, S.; Catizone, A.; et al. Microgravity Exposure Induces Antioxidant Barrier Deregulation and Mitochondrial Structure Alterations in TCam-2 Cells. *Biol. Life Sci. Forum* **2023**, *21*, 6. https://doi.org/10.3390/blsf2023021006

Academic Editor: Alexander E. Kalyuzhny

Published: 20 March 2023

Author Contributions: Conceptualization, G.R., L.G., A.C., M.A.M. and M.B.; methodology, L.G., M.B., A.R., M.Z., K.K., M.A.M., C.M. and S.G.; software, L.G., C.M., S.G. and M.Z.; validation, G.R., A.C., L.G. and M.B.; formal analysis, G.R., L.G., A.C. and M.B.; investigation, G.R., L.G., A.C. and M.B.; resources G.R., L.G., A.C. and M.B.; data curation, G.R., L.G., A.C., M.B., C.M. and A.R.; writing—original draft preparation, G.R. and A.C.; writing—review and editing, L.G. and M.B.; visualization G.R. and A.C.; supervision, F.F.; project administration, G.R., A.C. and F.F.; funding acquisition, A.C. and G.R. All authors have read and agreed to the published version of the manuscript.

Funding: This research was funded by ASI- Italian Space Agency grant number 2020-24-HH.0.

Institutional Review Board Statement: Not applicable.

Informed Consent Statement: Not applicable.

Data Availability Statement: Not applicable.

Conflicts of Interest: The authors declare no conflict of interest.

biology and life sciences forum

Abstract

Microgravity-Induced Metabolic Response in 2D and 3D TCam-2 Cell Cultures †

Caterina Morabito [1,*], Simone Guarnieri [1], Marika Berardini [2], Luisa Gesualdi [2], Francesca Ferranti [3], Anna Reale [4], Giulia Ricci [5], Angela Catizone [2] and Maria A. Mariggiò [1]

1 Department of Neuroscience, Imaging and Clinical Sciences-CAST, "G. d' Annunzio" University of Chieti-Pescara, 66100 Chieti, Italy; simone.guarnieri@unich.it (S.G.); maria.mariggio@unich.it (M.A.M.)
2 Department of Anatomy, Histology, Forensic-Medicine and Orthopedics, Section of Histology and Embryology, Sapienza University of Rome, 00185 Rome, Italy; marika.berardini@uniroma1.it (M.B.); luisa.gesualdi@uniroma1.it (L.G.); angela.catizone@uniroma1.it (A.C.)
3 Human Spaceflight and Scientific Research Unit, Italian Space Agency, 00133 Rome, Italy; francesca.ferranti@asi.it
4 Department of Experimental Medicine, Sapienza University of Rome, 00185 Rome, Italy; anna.reale@uniroma1.it
5 Department of Experimental Medicine, Università degli Studi della Campania Luigi Vanvitelli, 80138 Napoli, Italy; giulia.ricci@unicampania.it
* Correspondence: caterina.morabito@unich.it

† Presented at Cells, Cells and Nothing but Cells: Discoveries, Challenges and Directions, 6–8 March 2023; Available online: https://cells2023.sciforum.net/.

Abstract: The past few decades have seen an increasing number of both space travels and studies aimed at investigating the effects induced by space flights and the environment on humans. One of the main features of these conditions is the presence of altered gravity, mostly represented by microgravity experienced by astronauts. Microgravity is well known to induce deleterious effects at cellular, organ and systemic levels, including alterations in the male and female reproductive systems. In the present study, we investigated the effect of simulated microgravity on the metabolic activity of male germ cells using TCam-2 line as a cell model. These cells were cultured in the Random Positioning Machine that simulated microgravity conditions, and were grown as 2D monolayers or 3D spheroids to assay the effects on single cells or on organ-like structures. After a 24 hour-exposure to simulated microgravity, TCam-2 monolayers showed: (1) a decreased proliferation rate and a delay in cell cycle progression; (2) increased anaerobic metabolism; (3) increased levels of reactive oxygen species and superoxide anion; (4) modifications in mitochondrial morphology. After the same 24 hour-exposure, TCam-2 spheroids showed: (1) an increased anaerobic and aerobic activity in 40% and 26% of samples, respectively; (2) alterations in the redox balance with a decrease in catalase activity in about 65% of cell samples, and therefore, a deficit in the cellular antioxidant capacity; (3) increases in oxidative damage to proteins and lipids in more than 50% of cell samples. In conclusion, these data demonstrated a clear inference of simulated microgravity on the metabolic activity of TCam-2 cells, which is expressed through the activation of an oxidative stress state, that, if not compensated for, could be deleted over time.

Keywords: TCam-2 cells; cellular spheroids; simulated microgravity; ROS; oxidative stress; cellular metabolism

Citation: Morabito, C.; Guarnieri, S.; Berardini, M.; Gesualdi, L.; Ferranti, F.; Reale, A.; Ricci, G.; Catizone, A.; Mariggiò, M.A. Microgravity-Induced Metabolic Response in 2D and 3D TCam-2 Cell Cultures. *Biol. Life Sci. Forum* **2023**, *21*, 7. https://doi.org/10.3390/blsf2023021007

Academic Editor: Alexander E. Kalyuzhny

Published: 20 March 2023

Author Contributions: Conceptualization, C.M., G.R., A.C. and M.A.M.; methodology, C.M., S.G., M.B., L.G. and A.R.; software, C.M., M.B., L.G. and S.G.; validation, C.M., S.G. and M.A.M.; formal analysis, C.M., S.G. and M.A.M.; investigation, C.M., S.G. and M.A.M.; resources, A.C., G.R. and M.A.M.; data curation, C.M., S.G. and M.A.M.; writing—original draft preparation, C.M.; writing—review and editing, C.M., S.G., M.B., L.G., F.F., G.R., A.C. and M.A.M.; visualization, C.M. and

M.A.M.; supervision, M.A.M.; project administration, M.A.M. and F.F.; funding acquisition, G.R., A.C. and M.A.M. All authors have read and agreed to the published version of the manuscript.

Funding: This research was funded by ASI–Italian Space Agency, grant number 2020-24-HH.0.

Institutional Review Board Statement: Not applicable.

Informed Consent Statement: Not applicable.

Data Availability Statement: Not applicable.

Conflicts of Interest: The authors declare no conflict of interest.

biology and life sciences forum

Abstract

Metabolic Activity of *Chlamydomonas reinhardtii* Cells under Diclofenac-Induced Stress †

Darya Harshkova [1,*], Ivan Liakh [2], Pavel Hrouzek [3], Katerina Bisova [4], Bartosz Wielgomas [2] and Anna Aksmann [1]

1 Department of Plant Physiology and Biotechnology, Faculty of Biology, University of Gdansk, 80-308 Gdansk, Poland; anna.aksmann@ug.edu.pl
2 Department of Toxicology, Faculty of Pharmacy, Medical University of Gdansk, 80-416 Gdansk, Poland; ivan.liakh@gumed.edu.pl (I.L.); bartosz.wielgomas@gumed.edu.pl (B.W.)
3 Laboratory of Algal Biotechnology, Centre Algatech, Institute of Microbiology, Czech Academy of Sciences, 37981 Trebon, Czech Republic; hrouzek@alga.cz
4 Laboratory of Cell Cycles of Algae, Centre Algatech, Institute of Microbiology, Czech Academy of Sciences, 37981 Trebon, Czech Republic; bisova@alga.cz
* Correspondence: darya.harshkova@ug.edu.pl
† Presented at Cells, Cells and Nothing but Cells: Discoveries, Challenges and Directions, 6–8 March 2023; Available online: https://sciforum.net/event/cells2023.

Abstract: Non-steroidal anti-inflammatory drugs (NSAIDs), such as diclofenac (DCF), are detected in water bodies all over the world. Their presence in water environments pose a serious threat to non-target plant organisms, including unicellular green algae. To survive in the contaminated environments, these organisms need to modify their metabolism to be able to cope with NSAID-induced stress. Knowledge of the algal response to drugs is crucial for environmental protection. In the present work, we report the response of the unicellular green alga, *Chlamydomonas reinhardtii*, to DCF applied at a concentration of 32.7 mg/L, corresponding to toxicological parameter EC10. The algae's susceptibility to DCF was estimated based on the physiological parameters: population growth, oxidative stress symptoms, and photosynthetic activity. Moreover, the cell cultures were analyzed for the appearance of diclofenac transformation products. We found that DCF caused a slight decrease in the population growth rate and photosynthetic activity (quantum yield of photosynthesis) of the cells. Furthermore, some symptoms of oxidative stress (singlet oxygen overproduction) were observed. However, in the biomass and culture media, a wide range of DCF metabolites was discovered. This suggests that in the presence of relatively low concentrations of DCF, the biochemical activity of the algae was efficient enough to metabolize a part of the drug in the medium. Notably, some of the analyzed transformation products were similar to those formed during the metabolism of DCF by bacteria, while others were characteristic of eucaryotic metabolic pathways. In conclusion, *C. reinhardtii* exposed to DCF can keep its metabolic activity at a level sufficient for survival and biotransformation of the drug. Our results give rise to the assumption that other algae strains may also have the potential to metabolize DCF, thus contributing to the remediation of environments contaminated with pharmaceuticals.

Keywords: non-steroidal anti-inflammatory drugs; diclofenac; *Chlamydomonas reinhardtii*; metabolism

Citation: Harshkova, D.; Liakh, I.; Hrouzek, P.; Bisova, K.; Wielgomas, B.; Aksmann, A. Metabolic Activity of *Chlamydomonas reinhardtii* Cells under Diclofenac-Induced Stress. *Biol. Life Sci. Forum* **2023**, *21*, 8. https://doi.org/10.3390/blsf2023021008

Academic Editor: Alexander E. Kalyuzhny

Published: 20 March 2023

Author Contributions: Conceptualization, P.H. and B.W.; methodology, K.B. and P.H.; validation, K.B., P.H., A.A. and B.W.; formal analysis, D.H. and I.L.; investigation, D.H. and I.L.; resources, K.B. and P.H.; data curation, D.H. and I.L.; writing—original draft preparation, D.H.; writing—review and editing, I.L., A.A., K.B., P.H. and B.W.; visualization, D.H.; supervision, P.H., K.B. and A.A.; project administration, A.A.; funding acquisition, A.A. All authors have read and agreed to the published version of the manuscript.

Funding: This research was funded by National Science Centre Poland grant number UMO-2019/35/B/NZ9/01567.

Informed Consent Statement: Not applicable.

Data Availability Statement: Data will be available on the request.

Acknowledgments: The authors are grateful to ALGATECH—The Centre of Algal Biotechnology of the Institute of Microbiology of the Czech Academy Sciences (Czech Republic) for the support and methodological help.

Conflicts of Interest: The authors declare no conflict of interest.

biology and life sciences forum

MDPI

Abstract

The Effect of *RAS2* Gene Mutation in Single Cell Yeast Model †

Nazela Ahmed Chowdhury, Fatema Akter Sadia, Ismat Jahan Anee and Ashfaqul Muid Khandaker *

Branch of Genetics and Molecular Biology, Department of Zoology, University of Dhaka, Dhaka 1000, Bangladesh; chowdhurynazela@gmail.com (N.A.C.); fatemasadia03@gmail.com (F.A.S.); ismatjahan005@gmail.com (I.J.A.)

* Correspondence: muid.zoo@du.ac.bd; Tel.: +88-01759-044481

† Presented at Cells, Cells and Nothing but Cells: Discoveries, Challenges and Directions, 6–8 March 2023; Available online: https://sciforum.net/event/cells2023.

Abstract: More than 30% of all human malignancies are brought about by mutations in RAS proto-oncogenes (HRAS, KRAS, and NRAS) that are greatly conserved in yeast *RAS1* and *RAS2*. This makes yeast (*Saccharomyces cerevisiae*) an efficient single-celled eukaryotic model organism to study their functions. In the current investigation, the null mutation of the *RAS2* gene was analyzed to find out its deleterious consequences in yeast cells based on their ability to utilize glycerol as a respiratory substrate, mtDNA mutation rate, mtDNA abundance, and distribution pattern. Mutant cells grown in YPEG plates demonstrated slight respiratory deficiency compared to the wild type. An erythromycin-resistant assay was carried out to analyze the spontaneous mitochondrial DNA mutation rate in the Δ*ras2* mutant and it was found to be greater than that of wild type. In addition, the mitochondrial DNAs of both strains were also visualized under a fluorescence microscope using DAPI fluorescent stain. It was observed that mtDNA abundance was much lower compared to wild type cells. Thus, the present investigation revealed that deletion of the *RAS2* gene resulted in mtDNA mutation and depletion.

Keywords: mutation; RAS proto-oncogenes; *S. cerevisiae*; mitochondrial DNA; cancer

Citation: Chowdhury, N.A.; Sadia, F.A.; Anee, I.J.; Khandaker, A.M. The Effect of *RAS2* Gene Mutation in Single Cell Yeast Model. *Biol. Life Sci. Forum* **2023**, *21*, 9. https://doi.org/10.3390/blsf2023021009

Academic Editor: Alexander E. Kalyuzhny

Published: 20 March 2023

Author Contributions: Experimental work, N.A.C., F.A.S. and I.J.A.; writing—original draft preparation, N.A.C.; writing—review and editing, A.M.K.; supervision, A.M.K. All authors have read and agreed to the published version of the manuscript.

Funding: The work was funded by the Ministry of Science and Technology, Bangladesh.

Institutional Review Board Statement: Not applicable.

Informed Consent Statement: Not applicable.

Data Availability Statement: Not applicable.

Acknowledgments: The authors are greatly thankful to Niamul Naser for his kind support and assistance.

Conflicts of Interest: The authors declare no conflict of interest.

biology and life sciences forum

Abstract

Compartmentalization and Trafficking in Endoplasmic Reticulum Protein Quality Control †

Gerardo Z. Lederkremer [1,2]

[1] The Shmunis School of Biomedicine and Cancer Research, Tel Aviv University, Tel Aviv 69978, Israel; gerardol@tauex.tau.ac.il
[2] Sagol School of Neuroscience, Tel Aviv University, Tel Aviv 69978, Israel
† Presented at Cells, Cells and Nothing but Cells: Discoveries, Challenges and Directions, 6–8 March 2023; Available online: https://cells2023.sciforum.net/.

Abstract: Following translocation into the rough endoplasmic reticulum (ER), secretory proteins undergo a series of folding, maturation, compartmentalization and trafficking events. These are finely tuned to avoid misfolded protein accumulation and the consequent ER stress. Misfolded proteins and components of the ER quality control and ER-associated degradation (ERAD) machineries concentrate in mammalian cells in the pericentriolar ER-derived quality control compartment (ERQC), a staging ground for ERAD. We have recently determined that, surprisingly, trafficking to the ERQC and delivery to ERAD are dependent on COPII-coated vesicle transport and that they can be retrieved to the peripheral ER in COPI-coated vesicles.

Keywords: endoplasmic reticulum; ERAD; vesicular trafficking

Citation: Lederkremer, G.Z. Compartmentalization and Trafficking in Endoplasmic Reticulum Protein Quality Control. *Biol. Life Sci. Forum* **2023**, *21*, 10. https://doi.org/10.3390/blsf2023021010

Academic Editor: Alexander E. Kalyuzhny

Published: 20 March 2023

Funding: This research was funded by Israel Science Foundation grant numbers 1593/16 and 2577/20.

Institutional Review Board Statement: Not applicable.

Informed Consent Statement: Not applicable.

Data Availability Statement: Data available upon request.

Acknowledgments: I would like to thank members of the Lederkremer lab.

Conflicts of Interest: The author declares no conflict of interest.

biology and life sciences forum

MDPI

Abstract

Role of Gal-3 on Cisplatin-Induced Acute Liver Injury Model †

Diego Dias dos Santos [1,2,*], Nycole Morelli Belote [2], Rafael André da Silva [1], Adriana Aparecida Ferraz Carbonel [2], Gisela Rodrigues da Silva Sasso [2] and Cristiane Damas Gil [1,2]

1 Biosciences Graduate Program, Institute of Biosciences, Letters and Exact Sciences, São Paulo State University (IBILCE/UNESP), São José do Rio Preto 15054-000, SP, Brazil; rafaels@usp.br (R.A.d.S.); cristiane.gil@unifesp.br (C.D.G.)

2 Structural and Functional Biology Graduate Program, Paulista School of Medicine, Federal University of São Paulo (EPM/UNIFESP), São Paulo 04023-900, SP, Brazil; nycole.morelli@unifesp.br (N.M.B.); carbonel@unifesp.br (A.A.F.C.); gisela.morf@hotmail.com (G.R.d.S.S.)

* Correspondence: diego.dias@unesp.br; Tel.: +55-(11)-97656-9345

† Presented at Cells, Cells and Nothing but Cells: Discoveries, Challenges and Directions, 6–8 March 2023; Available online: https://cells2023.sciforum.net/.

Abstract: Oxidative stress is a common mechanism in the cytotoxicity of cisplatin, a widely used antineoplastic agent related to hepatotoxicity. In this context, we highlight galectin-3 (Gal-3), a β-galactoside-binding protein that regulates the inflammatory response and oxidative stress, and modified citrus pectin (MCP), an inhibitor of Gal-3. Thus, this study evaluates the effect of Gal-3 inhibition with MCP on cisplatin-induced acute liver injury in Wistar rats. Animals were divided into four groups (n = 5/group): SHAM–intraperitoneal (i.p.) injection of saline for 3 days; CIS–i.p. injection of cisplatin (10 mg/kg/day) for 3 days; MCP-orogastric gavage with MCP (100 mg/kg/day) for 7 days, followed by saline via i.p.; and MCP+CIS-gavage with MCP for 7 days, followed by cisplatin via i.p. for 3 days. Cisplatin administration caused a significant weight loss in the animals from CIS and MCP+CIS, an effect corroborated by a marked reduction in the glycogen storage in hepatocytes compared to their control groups. Cisplatin also provoked a marked increase in the influx of leukocytes, liver degeneration, ROS production, and STAT3 activation in the hepatocytes, plasma levels of cytokines (IL-6, IL-10), and hepatic toxicity biomarkers (ARG1, GSTα, SDH). Cisplatin per se reduced Gal-3 levels, especially in the mitochondria of hepatocytes. On the other hand, the MCP+CIS group also showed increased levels of IL-1β, TNF-α, and GOT1, as well as raised hepatic levels of MDA production and mitochondrial respiratory complex I. In conclusion, the inhibition of Gal-3 with MCP did not protect the liver against the deleterious effects of cisplatin, indicating that Gal-3 is important for tissue, cellular, and molecular maintenance of the liver.

Keywords: cytokines; hepatotoxicity; inflammation; mitochondria; modified citrus pectin; ROS

Citation: dos Santos, D.D.; Belote, N.M.; da Silva, R.A.; Carbonel, A.A.F.; da Silva Sasso, G.R.; Gil, C.D. Role of Gal-3 on Cisplatin-Induced Acute Liver Injury Model. *Biol. Life Sci. Forum* **2023**, *21*, 11. https://doi.org/10.3390/blsf2023021011

Academic Editor: Alexander E. Kalyuzhny

Published: 20 March 2023

Author Contributions: Conceptualization, D.D.d.S. and C.D.G.; methodology, D.D.d.S., N.M.B., R.A.d.S., A.A.F.C., G.R.d.S.S. and C.D.G.; formal analysis, D.D.d.S.; resources, C.D.G.; data curation, D.D.d.S., N.M.B. and C.D.G.; writing—original draft preparation, D.D.d.S. and C.D.G.; writing—review and editing, D.D.d.S. and C.D.G.; supervision, C.D.G.; project administration, C.D.G.; funding acquisition, C.D.G. All authors have read and agreed to the published version of the manuscript.

Funding: This research was funded by Fundação de Amparo à Pesquisa do Estado 9 de São Paulo—FAPESP [grant number 20/03565-2]. Diego Dias dos Santos is supported by CAPES 10 scholarship [code No. 001].

Institutional Review Board Statement: The experimental rat model was conducted according to the rules issued by the National Council for Control of Animal Experimentation (CONCEA) and approved by the Ethics Committee on Animal Use of the Federal University of São Paulo (CEUA/UNIFESP) in the meeting of 20 January 2021 (protocol code 5533211220).

Informed Consent Statement: Not applicable.

Data Availability Statement: Data will be made available upon request.

Conflicts of Interest: The authors declare no conflict of interest.

Abbreviations

ARG1	hepatic arginase 1
GOT1	aspartate transaminase 1
GSTα	α-glutathione S-transferase
IL	interleukin
MCP	modified citrus pectin
MDA	malondialdehyde
ROS	reactive oxygen species
SDH	sorbitol dehydrogenase
TNF-α	tumor necrosis factor-α
STAT3	signal transducer and activator of transcription 3

biology and life sciences forum

MDPI

Abstract

Annexin A1 Regulates Retinal Gliosis in Diabetic Retinopathy †

Rafael André da Silva [1,*], Luiz Philipe de Souza Ferreira [2], Vinicius M. Paiva Roda [3], Daniel R. Bastos [4] and Cristiane D. Gil [1,2]

1 Biosciences Graduate Program, Institute of Biosciences, Letters and Exact Sciences, São Paulo State University (IBILCE/UNESP), São José do Rio Preto 15054-000, SP, Brazil; cristiane.gil@unifesp.br
2 Structural and Functional Biology Graduate Program, Paulista School of Medicine, Federal University of São Paulo (EPM-UNIFESP), São Paulo 04023-900, SP, Brazil; luiz.philipe@unifesp.br
3 Life Systems Biology Graduate Program, Institute of Biomedical Sciences, Universidade de São Paulo (USP), São Paulo 05508-000, SP, Brazil; vinicius.moraes95@usp.br
4 Department of Oncology, Universidade de São Paulo, São Paulo 01246-903, SP, Brazil; danielbastos.adm@gmail.com
* Correspondence: rafael.andre@unesp.br

† Presented at Cells, Cells and Nothing but Cells: Discoveries, Challenges and Directions, 6–8 March 2023; Available online: https://cells2023.sciforum.net/.

Abstract: In diabetic retinopathy (DR), Müller cell gliosis contributes to retinal degeneration and inflammation. In this context, we highlight annexin A1 (AnxA1), an anti-inflammatory protein able to regulate neurodegeneration and angiogenesis; however, its mechanisms of action were poorly explored in DR. This study evaluates the function of AnxA1 in streptozotocin (STZ)-induced DR in wild-type (WT) and knockout (AnxA1$^{-/-}$) mice after 12 weeks. In addition, in silico analysis was performed with GSE111465 (whole retinas from 6-week-old STZ-diabetic or control animals) and GSE160306 (human retinas with different stages of DR). Retinas from 6-week-old STZ-diabetic mice showed raised transcripts of AnxA1 and GFAP compared to the controls. After 12 weeks, RD was associated with increased levels of AnxA1, formyl peptide receptor 2 (Fpr2) in the WT retina, as well as cleaved caspase 3 and vascular endothelial growth factor (VEGF) compared to the control samples. The lack of AnxA1 caused increased glutamine synthetase expression (Müller cell marker) in the retinas from RD animals compared to the WT RD group. On the other hand, no alterations in the levels of caspase 3 and VEGF expression were showed in the AnxA1$^{-/-}$ groups. Despite both genotypes presenting with gliosis in the peripheral retinas, as shown by glial fibrillary acid protein (GFAP) immunostaining, the AnxA1$^{-/-}$ RD group exhibited decreased levels of GFAP compared to the RD WT group. In an in silico study with human retinas, the severity of DR is associated with higher levels of AnxA1 mRNA expression. Additionally, a positive correlation between AnxA1 and GFAP mRNA levels was detected. These results allow us to conclude that AnxA1 participates in the progression of RD and that this protein can regulate the expression of GFAP.

Keywords: annexin A1; gliosis; Müller cells; diabetes; diabetic retinopathy; estreptozotocin; transcriptome

Citation: da Silva, R.A.; de Souza Ferreira, L.P.; Paiva Roda, V.M.; Bastos, D.R.; Gil, C.D. Annexin A1 Regulates Retinal Gliosis in Diabetic Retinopathy. *Biol. Life Sci. Forum* **2023**, *21*, 12. https://doi.org/10.3390/blsf2023021012

Academic Editor: Alexander E. Kalyuzhny

Published: 20 March 2023

Author Contributions: Conceptualization R.A.d.S.; methodology: R.A.d.S. and C.D.G.; formal analysis and investigation: R.A.d.S. and C.D.G.; experimental procedures: R.A.d.S., L.P.d.S.F., V.M.P.R., D.R.B. and C.D.G.; funding acquisition: C.D.G.; resources: C.D.G.; supervision: C.D.G. All authors have read and agreed to the published version of the manuscript.

Funding: Fundação de Amparo à Pesquisa do Estado de São Paulo (FAPESP), grant number 2022/02327-6. R.A.S., V.M.P.R and D.R.B was supported by the Coordenação de Aperfeiçoamento de Pessoal de Nível Superior (CAPES), finance code 001. L.P.S.F was supported by the FAPESP, grant number 2021/00270-4.

Institutional Review Board Statement: The experimental mouse model was conducted according to the Brazilian Law 11.794 of 8 October 2008, Decree 6899 of 15 July 2009, as well as with the rules issued by the National Council for Control of Animal Experimentation (CONCEA) and approved by the Ethics Committee on Animal Use of the Federal University of São Paulo (CEUA/UNIFESP) in the meeting of 08/09/2021 (protocol code 8518230821).

Informed Consent Statement: Not applicable.

Data Availability Statement: Data available on request due to restrictions.

Conflicts of Interest: The authors declare no conflict of interest.

biology and life sciences forum

Abstract

The Senescence Marker p16Ink4a—A Player of Liver Endothelial Cells Physiology †

Hasan Safwan-Zaiter *, Kay-Dietrich Wagner * and Nicole Wagner *

CNRS, INSERM, iBV, Université Côte d'Azur, 06107 Nice, France

* Correspondence: hasan.safwan-zaiter@unice.fr (H.S.-Z.); kwagner@unice.fr (K.-D.W.); nwagner@unice.fr (N.W.)

† Presented at Cells, Cells and Nothing but Cells: Discoveries, Challenges and Directions, 6–8 March 2023; Available online: https://cells2023.sciforum.net/.

Abstract: P16INK4A is a tumor suppressor and cell cycle regulator that has been linked to aging and senescence. In development, a potential role of p21 and of p19ARF has been postulated, but little is known about p16. Our previous results revealed a highly dynamic expression pattern of p16 in development and in different organs and cell types assessed by qRT-PCR and immunohistochemistry (IHC). In addition, we also noticed through IHC observation that p16 expression in old liver is mainly in the endothelial cells (ECs) compared to parenchymal cells. Therefore, we aimed at better understanding the role of p16 in biological processes of liver ECs, such as proliferation, migration, apoptosis, and tube formation. We also performed RNA sequencing to identify genes differentially expressed between young and old ECs. We used small hairpin (shRNA) constructs and a p16 cDNA-GFP vector to knockdown and overexpress p16 in vitro, in two types of liver ECs, CD31+ vascular ECs and CD146+ sinusoidal endothelial cells. Afterwards, we assessed p16's down- and up-regulation effect on ECs function. Brdu incorporation assays showed that p16 upregulation was associated with slower proliferation compared to control cells, whereas its downregulation induced higher proliferation compared to control cells. Scratch assay and transwell migration assays showed attenuated migration in p16 overexpressed cells compared to baseline expression, while only transwell assays showed the ameliorated migration of p16 knockdown cells compared to controls. Similar migration between p16 knockdown and control was observed in scratch assays. We also observed in β-gal staining, a marker of senescence, a higher number of stained cells in p16 overexpression conditions compared to controls, while less cells were stained in the case of knockdown. Additional experiments that aim to further decipher p16's effect in ECs' tube formation, apoptosis, and telomeres shortening are ongoing, which might contribute to the invention of more specialized anti-aging therapies.

Keywords: aging; development; endothelial cells; liver; p16; senescence

Citation: Safwan-Zaiter, H.; Wagner, K.-D.; Wagner, N. The Senescence Marker p16Ink4a—A Player of Liver Endothelial Cells Physiology. *Biol. Life Sci. Forum* **2023**, *21*, 13. https://doi.org/10.3390/blsf2023021013

Academic Editor: Alexander E. Kalyuzhny

Published: 20 March 2023

Author Contributions: Conceptualization, N.W. and K.-D.W.; methodology, H.S.-Z., N.W. and K.-D.W.; formal analysis, H.S.-Z., N.W. and K.-D.W.; investigation, H.S.-Z., N.W. and K.-D.W.; resources, N.W. and K.-D.W.; writing—original draft preparation, H.S.-Z., N.W. and K.-D.W.; writing—review and editing, N.W. and K.-D.W.; visualization, H.S.-Z. and N.W.; supervision, N.W. and K.-D.W.; project administration, N.W. and K.-D.W.; funding acquisition, N.W. and K.-D.W. All authors have read and agreed to the published version of the manuscript.

Funding: This research was funded by Fondation pour la Recherche Medicale, grant number FRM DPC20170139474 (K.-D.W.), Fondation ARC pour la recherche sur le cancer, grant number PJA 20161204650 (N.W.), Gemluc (N.W.), Plan Cancer INSERM (K.-D.W.), and Agence Nationale de la Recherche, grant R19125AA "Senage" (K.-D.W.).

Institutional Review Board Statement: Not applicable.

Informed Consent Statement: Not applicable.

Data Availability Statement: Not applicable.

Conflicts of Interest: The authors declare no conflict of interest.

biology and life sciences forum

MDPI

Abstract

An Unveiled Cell Death Mechanism Exclusive to Human Cancer Cells †

Malka Cohen-Armon

Sackler School of Medicine and Sagol School of Neuroscience, Tel-Aviv University, Tel-Aviv 69978, Israel; marmon@tauex.tau.ac.il

† Presented at Cells, Cells and Nothing but Cells: Discoveries, Challenges and Directions, 6–8 March 2023; Available online: https://cells2023.sciforum.net/.

Abstract: The modified phenanthridine PJ34 blocks the post-translational modifications of specific proteins highly expressed in human malignant cells. This exclusively arrests mitosis in human malignant cells by inserting flaws in their mitotic spindle structure. Cancer cells were efficiently eradicated by Mitotic Catastrophe cell death, while similarly treated healthy proliferating cells were spared and continued to proliferate as untreated cells. This cytotoxic effect was examined in a variety of human epithelial cancer cells in tissue culture and in xenografts. Three affected proteins were identified out of all tested proteins implicated in mitosis in epithelial malignant cells compared to healthy epithelial cells. Two kinesins, KifC1/HSET and Kif18A, as well as NuMA were identified. The identified kinesins are already examined for their potential implication in cancer therapy. Blocking the post-translational modifications of NuMA by PJ34 exclusively prevented the protein binding capacity of NuMA in cancer cells. This prevented its clustering in the spindle poles, which stabilizes the spindles and enables alignment of chromosomes in spindle mid-zone. Un-aligned chromosomes and dispersed NuMA and centrosomes were detected in distorted spindles of human cancer cells treated with PJ34. Mitosis was arrested in the anaphase and this lead to cell death via cytochrome-c leakage from the mitochondria membrane. Thus, the cytotoxic activity of PJ34 unveiled a new mechanism causing self-eradication of human cancer cells during mitosis, including cancer cells that are not responsive to current therapies and regardless of specific mutations. The more rapidly cells proliferate, the more rapidly they are eradicated.

Keywords: mitotic spindle; cancer cells; NuMA; kinesins; Mitotic Catastrophe cell death

Citation: Cohen-Armon, M. An Unveiled Cell Death Mechanism Exclusive to Human Cancer Cells. *Biol. Life Sci. Forum* **2023**, *21*, 14. https://doi.org/10.3390/blsf2023021014

Academic Editor: Alexander E. Kalyuzhny

Published: 20 March 2023

Funding: This research was funded by Novartis.

Institutional Review Board Statement: Not applicable.

Informed Consent Statement: Not applicable.

Data Availability Statement: The data is available in [1].

Conflicts of Interest: The author declares no conflict of interest.

Reference

1. Cohen-Armon, M. Exclusive modifications of NuMA in malignant epithelial cells: A potential therapeutic mechanism. *Drug. Dis. Today* **2022**, *27*, 1205–1209. [CrossRef] [PubMed]

biology and life sciences forum

MDPI

Abstract

Untangling the Microscopic World of Organelles, Cells, Tissues, and Organs: A Focus on the Dysfunctional Golgi Apparatus in Disease Research †

Daniel Joseph Gómez [1,2]

[1] SLAC National Accelerator Laboratory, Stanford University, Menlo Park, CA 94025, USA; dgomez@slac.stanford.edu

[2] Department of Biological Sciences, California State University, East Bay, Hayward, CA 94542, USA

† Presented at Cells, Cells and Nothing but Cells: Discoveries, Challenges and Directions, 6–8 March 2023; Available online: https://cells2023.sciforum.net/.

Abstract: Emerging techniques in organelle structural biology have revolutionized our understanding of disease mechanisms and opened new possibilities for developing targeted therapies. In particular, dysfunctions of the Golgi apparatus (GA) have been implicated in a wide range of neurological disorders and cancer, making it a key area of focus in organelle structural biology. The GA plays a crucial role in regulating the transport and modification of proteins and lipids, and dysfunction of this organelle can lead to mislocation and accumulation of proteins and impaired glycosylation, resulting in neurodegenerative diseases such as Parkinson's Disease and neurodevelopmental disorders (NDDs). Inhibition of vesicular trafficking by α-synuclein may affect dopamine-producing neurons and neuromodulators, while fragmentation and defects within the GA can lead to apoptotic pathways during pathological mechanisms. Additionally, defects and fragmentation of the GA have been implicated in cancer progression, making it a key area of interest for cancer researchers. Advances in imaging technology, such as cryogenic electron tomography, soft-X-ray tomography (SXT), and multiplex correlative light and electron microscopy (CLEM), have enabled high-resolution visualization of the GA and its dysfunctions in neurological diseases and cancer. These techniques provide detailed insight into the structure and function of the GA and have the potential to inform new treatments for diseases associated with GA dysfunction. Recent studies have shown that molecular zippers hold the Golgi membrane together, providing further insight into the mechanisms underlying GA dysfunction in diseases such as Parkinson's, NDDs, and cancer. Cryo-CLEM and nanobody-assisted tissue immunostaining for volumetric EM (NATIVE) techniques enable high-resolution visualization of the GA and its native environment, aiding in understanding its function in health and disease. In addition, novel techniques such as Optical coherence tomography (OCT) enable rapid, accurate, and high-resolution in vivo imaging of the mouse cortex, providing 3D visualization of cortical microarchitecture using a feature segmentation algorithm. OCT enables label-free, micron-scale 3D imaging of biological tissues' fine structures with significant depth and a large field of view. A 3D CNS segmentation mask of brain neural networks in a living mouse can be visualized at micron-level resolution using OCT. Overall, the organelle structural biology field, specifically the study of the Golgi apparatus dysfunction in neurological disorders and cancer, has significant implications for developing new therapeutic targets, gene therapy, and drug design. With continued research and advancements in imaging technologies, we can expect to gain a more comprehensive understanding of the underlying mechanisms of GA dysfunction in neurological disorders and cancer, paving the way for innovative new treatments and therapies.

Keywords: organelle structural biology; Golgi apparatus; neurodegenerative diseases; cancer; imaging technologies; cryogenic electron tomography; soft-X-ray tomography; optical coherence tomography; therapeutic targets; drug design

Citation: Gómez, D.J. Untangling the Microscopic World of Organelles, Cells, Tissues, and Organs: A Focus on the Dysfunctional Golgi Apparatus in Disease Research. *Biol. Life Sci. Forum* **2023**, *21*, 15. https://doi.org/10.3390/blsf2023021015

Academic Editor: Alexander E. Kalyuzhny

Published: 21 March 2023

Funding: This research received no external funding.

Institutional Review Board Statement: Not applicable.

Informed Consent Statement: Not applicable.

Data Availability Statement: The data is available at https://doi.org/10.20944/preprints202210.0383.v2 (accessed on 17 March 2023).

Acknowledgments: D.J.G. thanks members of the SSRL, SLAC, and LCLS and associated laboratories for helpful discussions.

Conflicts of Interest: D.J.G. is the Director of Multiomics in the Multiomics Structural Biology and Network Physiology Division, Neuro-Spatial Multi-Omics Structural Biology (NSMOSB) unit at Gomera Research. As the Lead Molecular Research Scientist and Biomedical AI/ML Engineer at Gomera, D.J.G. works on spatial multi-omics, quantum computational medicine, and machine learning. Additionally, D.J.G. is the founder and has equity in Gomera Health, Inc. and GomeBio LLC, and serves as a scientific advisory board member for Gomera Health, Inc.

biology and life sciences forum

MDPI

Abstract

Building Implantable Human Liver Tissue from Pluripotent Stem Cells †

David Hay

Centre for Regenerative Medicine, University of Edinburgh, Edinburgh EH16 4UU, UK; davehay@talktalk.net

† Presented at Cells, Cells and Nothing but Cells: Discoveries, Challenges and Directions, 6–8 March 2023; Available online: https://cells2023.sciforum.net/.

Abstract: Liver disease is an escalating global health issue. While liver transplantation is an effective mode of therapy, patient mortality has increased due to the shortage of donor organs. Therefore, developing renewable sources of human liver tissue is attractive. Pluripotent stem cell-derived liver tissue represents a potential alternative to cadaver derived hepatocytes and whole organ transplant. At present, two-dimensional differentiation procedures deliver tissue lacking certain functions and phenotypic instability. Efforts to overcome these limiting factors have led to the building of three-dimensional (3D) cellular aggregates. Although enabling for the field, their widespread application and adoption is limited due to the reliance on variable biological components. Our study focus on developing 3D liver tissue under defined conditions. We demonstrate that 3D derived tissue can be generated at scale and implanted underneath the skin of mice. Excitingly, implanted human tissue provides support to mice with metabolic liver disease. This includes immunocompetent recipients. In addition to their clinical application, in vitro generated 3D tissues have important roles to perform in developing safe and efficacious medicines to treat human diseases. We demonstrate that stem cell derived 3D liver tissue exhibits liver cell phenotype for over one year in culture, providing an attractive resource for long-term disease modelling and screening studies. In conclusion, stem cell derived liver tissue has great potential for in vitro and in vivo endeavours. Our most recent advances will be presented at the meeting.

Keywords: hESC; iPSC; liver; hepatocyte; endothelial cell; tissue engineering

Citation: Hay, D. Building Implantable Human Liver Tissue from Pluripotent Stem Cells. *Biol. Life Sci. Forum* **2023**, *21*, 16. https://doi.org/10.3390/blsf2023021016

Academic Editor: Alexander E. Kalyuzhny

Published: 21 March 2023

Funding: UK Regenerative Medicine Platform (MRC MR/L022974/1), Chief Scientist's Office (TCS/16/37).

Institutional Review Board Statement: Not applicable.

Informed Consent Statement: Not applicable.

Data Availability Statement: Supporting data can be found at https://link.springer.com/article/10.1007/s00204-018-2280-2 and https://iopscience.iop.org/article/10.1088/1758-5090/abbdb2 (accessed on 14 March 2023).

Conflicts of Interest: David Hay is a founder, director and shareholder in Stimuliver ApS and Stemnovate Limited.

biology and life sciences forum

MDPI

Abstract

CAVPENET Decreases Prostate Cancer Cells Proliferation and Invasion through Modulation of Protein Phosphatase Activity †

Bárbara Matos [1,2], John Howl [3], Carmen Jerónimo [2,4] and Margarida Fardilha [1,*]

1 Laboratory of Signal Transduction, Department of Medical Sciences, Institute of Biomedicine iBiMED, University of Aveiro, 3810-193 Aveiro, Portugal; barbaracostamatos@ua.pt

2 Cancer Biology and Epigenetics Group, IPO Porto Research Center (CI-IPOP), Portuguese Institute of Oncology of Porto (IPO Porto), Research Center-LAB 3, F Bdg., 1st floor, Rua Dr. António Bernardino de Almeida, 4200-072 Porto, Portugal; carmenjeronimo@ipoporto.min-saude.pt

3 Molecular Pharmacology Group, Research Institute in Healthcare Science, University of Wolverhampton, Wolverhampton WV1 1LY, UK; j.howl@wlv.ac.uk

4 Department of Pathology and Molecular Immunology, Institute of Biomedical Sciences Abel Salazar, University of Porto (ICBAS-UP), Rua Jorge Viterbo Ferreira 228, 4050-513 Porto, Portugal

* Correspondence: mfardilha@ua.pt

† Presented at the Cells, Cells and Nothing but Cells: Discoveries, Challenges and Directions, 6–8 March 2023; Available online: https://cells2023.sciforum.net/.

Abstract: Despite advances in understanding the molecular mechanisms underlying prostate cancer progression, the development of effective therapeutic approaches remains a major challenge. In this context, the protein phosphatase 1 (PP1) and its complexes have been recognized as potential drug targets. Herein, we designed and synthetized a peptide sequence based on the PP1-binding motif of CAV1, which was coupled with penetratin to improve cellular uptake. To evaluate the effect of the synthetized peptide, named CAVPENET, prostate cancer cells (PC-3 and LnCaP) were incubated with CAVPENET for 48 h, and several parameters were analyzed. We found that CAVPENET significantly decreased the LnCaP and PC-3 cells viability and invasive ability. A significant decrease in the phosphorylation of AKT at Ser^{473} was also observed after 48 h of incubation with CAVPENET. Moreover, a slight recovery of AKT phosphorylation levels after the simultaneous incubation of CAVPENET (10 μM) with tautomycin (10 nM)—a highly specific PP1 inhibitor—suggested the role of PP1 in the CAVPENET-induced alterations in AKT phosphorylation. Moreover, incubation with CAVPENET (10 μM) + cantharidin (0.5 μM), a potent and selective PP2A inhibitor, almost completely recovered the phosphorylation levels of AKT, suggesting the role of PP2A in the effect of CAVPENET. Altogether, these results highlight the potential of the synthesized peptide to negatively impact the PCa cells' proliferation and invasive ability by interfering with the interaction of CAV1 with PP1 and/or PP2A. Further analyses are now required to confirm the disruption of the interactions and to better elucidate the mechanisms of cell death.

Keywords: prostate cancer; PP1; PP2A; peptide; phosphatase; CAV1

Citation: Matos, B.; Howl, J.; Jerónimo, C.; Fardilha, M. CAVPENET Decreases Prostate Cancer Cells Proliferation and Invasion through Modulation of Protein Phosphatase Activity. *Biol. Life Sci. Forum* **2023**, *21*, 17. https://doi.org/10.3390/blsf2023021017

Academic Editor: Alexander E. Kalyuzhny

Published: 22 March 2023

Author Contributions: Conceptualization, B.M. and M.F.; methodology, B.M.; formal analysis, B.M. and M.F.; writing—original draft preparation, B.M.; writing—review and editing, J.H., C.J. and M.F.; supervision, J.H., C.J. and M.F.; funding acquisition, M.F. All authors have read and agreed to the published version of the manuscript.

Funding: This research was funded by the Institute for Biomedicine (iBiMED) (UIDB/04501/2020) and an individual scholarship for BM (SFRH/BD/146032/2019).

Institutional Review Board Statement: Not applicable.

Informed Consent Statement: Not applicable.

Data Availability Statement: Not applicable.

Conflicts of Interest: The authors declare no conflict of interest.

biology and life sciences forum

Abstract

Mechanistic Insights on the Anticancer Effects of Metformin in Primary Breast Cancer Cells †

Francesca Cirillo [1,*], Domenica Scordamaglia [1], Marianna Talia [1], Maria Francesca Santolla [1], Lucia Muglia [1], Azzurra Zicarelli [1], Salvatore De Rosis [1], Asia Spinelli [1], Francesca Giordano [1], Anna Maria Miglietta [2], Marcello Maggiolini [1] and Rosamaria Lappano [1]

[1] Department of Pharmacy, Health and Nutritional Sciences, University of Calabria, 87036 Rende, Italy; domenica.scordamaglia@unical.it (D.S.); marianna.talia@unical.it (M.T.); mariafrancesca.santolla@unical.it (M.F.S.); lucia.muglia@unical.it (L.M.); azzurra.zicarelli@unical.it (A.Z.); salvatore.derosis@libero.it (S.D.R.); asia.spinelli@unical.it (A.S.); francesca.giordano@unical.it (F.G.); marcello.maggiolini@unical.it (M.M.); rosamaria.lappano@unical.it (R.L.)

[2] Breast Unit, Regional Hospital Cosenza, 87100 Cosenza, Italy; annamariamiglietta@virgilio.it

* Correspondence: francesca.cirillo@unical.it

† Presented at Cells, Cells and Nothing but Cells: Discoveries, Challenges and Directions, 6–8 March 2023; Available online: https://cells2023.sciforum.net/.

Abstract: Metabolic disorders, such as obesity, type 2 diabetes (T2D) and metabolic syndrome, have been implicated in breast cancer (BC) progression. In this regard, insulin has been shown to promote mitogenic and metastatic responses in BC through diverse signaling pathways. Moreover, high levels of insulin and elevated expression of its cognate receptor, namely insulin receptor (IR), have been associated with increased BC incidence, resistance to treatments and poor outcomes. Metformin (1,1-dimethylbiguanide hydrochloride) is the most commonly prescribed drug for T2D treatment worldwide. Metformin has been shown to interfere with BC cell growth. In order to provide novel insights through which metformin can elicit anti-cancer responses in BC, we performed bioinformatics analysis as well as TaqMan Gene Expression Assay, flow cytometry, immunofluorescence, immunoblots, 2D and 3D proliferation assays and motility experiments. A naturally immortalized BC cell line (namely BCAHC-1) and important components of the tumor microenvironment, such as cancer-associated fibroblasts (CAFs) derived from BC patients, were used as model systems. We found that metformin inhibits the activation of main transduction pathways, the gene expression changes and the proliferative effects induced by insulin in BCAHC-1 cells. Moreover, metformin prevented the insulin-stimulated induction of CXC chemokine receptor 4 (CXCR4), which has been involved in BC metastatic dissemination. Next, metformin suppressed the invasion of CAFs triggered through CXCR4 via insulin stimulated BCAHC-1 cells. Our findings may suggest novel transduction mechanisms involved in the inhibitory effects elicited by metformin in both BC cells and CAFs.

Keywords: breast cancer; metformin; insulin/insulin receptor; tumor microenvironment

Citation: Cirillo, F.; Scordamaglia, D.; Talia, M.; Santolla, M.F.; Muglia, L.; Zicarelli, A.; De Rosis, S.; Spinelli, A.; Giordano, F.; Miglietta, A.M.; et al. Mechanistic Insights on the Anticancer Effects of Metformin in Primary Breast Cancer Cells. *Biol. Life Sci. Forum* **2023**, *21*, 18. https://doi.org/10.3390/blsf2023021018

Academic Editor: Alexander E. Kalyuzhny

Published: 22 March 2023

Author Contributions: Conceptualization, F.C., D.S., R.L. and M.M.; methodology, F.C., D.S., M.T., R.L. and M.M.; software, F.G., F.C. and M.T.; validation, F.C., D.S., M.T. and R.L.; formal analysis, M.T.; investigation, F.C., D.S., M.T., M.F.S., L.M., A.Z., S.D.R. and A.S.; resources, A.M.M. and M.M.; data curation, F.C., D.S. and M.T.; writing—original draft preparation, F.C. and D.S.; writing—review and editing, R.L. and M.M.; visualization R.L. and M.M.; supervision, R.L. and M.M.; project administration, M.M.; funding acquisition, M.M. All authors have read and agreed to the published version of the manuscript.

Funding: Fondazione AIRC supported MM (IG n. 21322) and RL (IG n. 27386). Ministero della Salute (Italy) supported MM and RL (project RF-2019-12368937).

Institutional Review Board Statement: The study was conducted in accordance with the Declaration of Helsinki for studies involving humans.

Informed Consent Statement: Written informed consent has been obtained from the patients to publish this paper and the experimental research has been performed with the ethical approval provided by the "Comitato Etico Ospedale Regionale, Cosenza, Italy" (approval code: 166, 2 December 2016).

Data Availability Statement: Not applicable.

Acknowledgments: We would like to thank the special award, namely "Department of Excellence 2018–2022" (Italian Law 232/2016) at the Department of Pharmacy, Health and Nutritional Sciences of the University of Calabria (Italy) and the "Sistema Integrato di Laboratori per L'Ambiente—(SILA) PONa3_00341".

Conflicts of Interest: The authors declare no conflict of interest.

biology and life sciences forum

MDPI

Abstract

Implication of Intra-Tumor Heterogeneity on Colorectal Cancer Response to MEK Inhibition †

Christian Regenbrecht [1,2,3]

[1] CELLphenomics GmbH, Robert-Rössle-Str. 10, 13125 Berlin, Germany; christian.regenbrecht@cellphenomics.com

[2] ASC Oncology GmbH, Robert-Rössle-Str. 10, 13125 Berlin, Germany

[3] Institute for Pathology, Robert-Koch-Str. 40, 37075 Göttingen, Germany

† Presented at Cells, Cells and Nothing but Cells: Discoveries, Challenges and Directions, 6–8 March 2023; Available online: https://cells2023.sciforum.net/.

Abstract: Intra-tumor heterogeneity (ITH) poses a major obstacle in cancer therapy. In colorectal cancer (CRC), mutations in the transforming growth factor-β/bone morphogenetic protein (TGF-β/BMP) pathway, especially in the SMAD4 gene, have been correlated with decreased overall survival and are suspected to modulate chemoresistance. We have previously shown that $SMAD4^{R361H}$ is associated with differential drug response towards EGFR, MEK and PI3K inhibitors. Here, we analyzed the mechanistic role of $SMAD4^{R361H}$ using oncoproteomics in CRISPR-engineered $SMAD4^{R361H}$ and CRC patient-derived organoids (PD3D®). Utilizing DigiWest® multiplex protein profiling analysis, we confirmed a stronger response to MEK inhibition in organoids harboring $SMAD4^{R361H}$ as compared to $SMAD4^{wt}$ PD3D. After 24 h of incubation with 0.03 μM trametinib, we observed a more pronounced decrease in proliferation markers, such as cyclin B1 and aurora kinase A in $SMAD4^{R361H}$ cells. Interestingly, there was no noticeable accumulation of caspases 3 and 9 in any organoid culture; however, there was a conspicuous trend in the accumulation of Bcl-xL in presence of $SMAD4^{R361H}$. To understand the underlying mechanism of such a discrepancy, we analyzed the protein levels and phosphorylation status of other SMADs, as $SMAD4^{R361H}$ disrupts TGF-β/BMP signal transduction. Out of all SMADs, only SMAD5 showed significant changes in protein level and phosphorylation status in response to the treatment only in $SMAD4^{wt}$ organoids. As previously published, BMP signaling promotes cancer cell proliferation and tumor growth. It is plausible to assume that functional loss of SMAD4 and thus loss of SMAD5 signaling renders the $SMAD4^{R361H}$ subpopulation of cells more sensitive to MEK inhibitors. Loss of SMAD4 was previously shown to promote chemoresistance and was associated with a higher recurrence rate in colorectal cancer. The heterogenic landscape of mutated SMAD4 within the same tumor, in this light, can give rise to multi-drug resistant disease.

Keywords: intra-tumor heterogeneity; cancer; cellular pathology

Citation: Regenbrecht, C. Implication of Intra-Tumor Heterogeneity on Colorectal Cancer Response to MEK Inhibition. *Biol. Life Sci. Forum* **2023**, *21*, 19. https://doi.org/10.3390/blsf2023021019

Academic Editor: Alexander E. Kalyuzhny

Published: 22 March 2023

Funding: This research received no external funding.

Institutional Review Board Statement: Not applicable.

Informed Consent Statement: Not applicable.

Data Availability Statement: Not applicable.

Conflicts of Interest: The author declares no conflict of interest.

biology and life sciences forum

Abstract

The Effect of the Synergistic Combination of Vitamin D and Doxorubicin on the MCF-7 Line Breast Cancer Cells †

Kamila Sobczak [1,*], Agnieszka Mlicka [1], Paweł Niewiadomski [1], Oliwia Abramczyk [1], Wioletta Zielińska [2], Marta Hałas-Wiśniewska [2] and Magdalena Izdebska [2]

1 Students Research Group of Cell Biology and Ultrastructure, Department of Histology and Embryology, Ludwik Rydygier Collegium Medicum in Bydgoszcz, Nicolaus Copernicus University in Toruń, 85-092 Bydgoszcz, Poland; a.aganaw@gmail.com (A.M.); nomski11@gmail.com (P.N.); oliwia.abramczyk@onet.pl (O.A.)

2 Department of Histology and Embryology, Faculty of Medicine, Ludwik Rydygier Collegium Medicum in Bydgoszcz, Nicolaus Copernicus University in Toruń, 85-092 Bydgoszcz, Poland; w.zielinska@cm.umk.pl (W.Z.); mhalas@cm.umk.pl (M.H.-W.); mizdebska@cm.umk.pl (M.I.)

* Correspondence: kamilas77533@gmail.com

† Presented at Cells, Cells and Nothing but Cells: Discoveries, Challenges and Directions, 6–8 March 2023; Available online: https://cells2023.sciforum.net/.

Abstract: Breast cancer is the most prevalent cancer in the female population. The prolonged action of estrogen may affect tumor proliferation. Additionally, a fat-rich diet may have various effects on cancer proliferation, depending on the type of fat. Vitamin D, similarly to estrogen, is a fat-soluble cholesterol derivative. The deficiency of vitamin D correlates with an increased proliferation of breast cancer cells. In turn, doxorubicin is commonly used as a cytostatic in chemotherapy. The study aimed to assess whether vitamin D enhances the anticancer effect of doxorubicin (DOX) in the MCF-7 cell line. The cells were divided into four groups: untreated control, DOX- and vitamin D-treated cells, and cells treated with the combination of compounds in a 1:1 ratio. We applied the MTT colorimetric assay (cell viability analysis), Annexin V/PI assay (cell death analysis), flow cytometry (cell cycle distribution), and fluorescence staining of cytoskeletal proteins (F-actin and vimentin). The type of DOX and vitamin D interaction was estimated based on the Chou–Talalay method. Our results showed that vitamin D and doxorubicin in a 1:1 ratio act synergistically. We observed a decrease in the survival of MCF7 cells. The combination of DOX and vitamin D enhanced the changes in morphology and organization in F-actin and the vimentin network compared to the treatment with the substances separately. In summary, we suggest that natural compounds such as vitamin D may be useful in anticancer treatment in the context of enhancing the cytostatic effects of drugs.

Keywords: breast cancer; vitamin D; doxorubicin

Citation: Sobczak, K.; Mlicka, A.; Niewiadomski, P.; Abramczyk, O.; Zielińska, W.; Hałas-Wiśniewska, M.; Izdebska, M. The Effect of the Synergistic Combination of Vitamin D and Doxorubicin on the MCF-7 Line Breast Cancer Cells. *Biol. Life Sci. Forum* **2023**, *21*, 20. https://doi.org/10.3390/blsf2023021020

Academic Editor: Alexander E. Kalyuzhny

Published: 22 March 2023

Author Contributions: Conceptualization, A.M.; methodology, K.S., O.A. and P.N.; validation, M.I.; formal analysis, M.H.-W.; data curation, W.Z.; writing—original draft preparation, W.Z., K.S. and O.A.; writing—review and editing, W.Z.; visualization, P.N. and M.H.-W.; supervision, M.I.; project administration, M.I.; funding acquisition, A.M. All authors have read and agreed to the published version of the manuscript.

Funding: This study was supported by research task within the framework of the Students Researches and a research task within the framework of the Department of Histology and Embryology (Nicolaus Copernicus University in Torun, Faculty of Medicine, Collegium Medicum in Bydgoszcz).

Institutional Review Board Statement: Not applicable.

Data Availability Statement: Not applicable.

Conflicts of Interest: The authors declare no conflict of interest.

Disclaimer/Publisher's Note: The statements, opinions and data contained in all publications are solely those of the individual author(s) and contributor(s) and not of MDPI and/or the editor(s). MDPI and/or the editor(s) disclaim responsibility for any injury to people or property resulting from any ideas, methods, instructions or products referred to in the content.

biology and life sciences forum

MDPI

Abstract

Huge Modification of the Cell Theory by the Recent Discovery of the Widespread Cell-Derived Extracellular Vesicles †

Irène Tatischeff

Honorary CNRS and UPMC Research Director, Founder of RevInterCell, a Scientific Consulting Service, 91400 Orsay, France; irene.tatischeff@upmc.fr

† Presented at Cells, Cells and Nothing but Cells: Discoveries, Challenges and Directions, 6–8 March 2023; Available online: https://cells2023.sciforum.net/.

Abstract: The aim of this work is to discuss the necessity to strongly modify the powerful well-acknowledged cell theory by taking into account the recently discovered universal cell-derived extracellular vesicles (EVs). In a great breakthrough, EVs are now known to mediate important cell interconnections, with many still unknown mechanisms. There is a missing step between the accumulated biological knowledge about EVs during the past two decades and the many recent preclinical searches, dealing with a few human patients compared to controls, for the applications of EVs in oncology. In this case, the huge amount of different cell-derived EVs generates an inextricable complexity. To evidence unknown EV-mediated mechanisms, a simple cell model would be much more convenient. The microorganism *Dictyostelium discoideum* (Dd) is ideal to achieve this goal as a wonderful eukaryotic in vitro and in vivo cell model. In 1998, we discovered Dd EVs to be involved in mediating a new multidrug resistance mechanism, and also the normal and physiological release of different EVs during the well-separated growth and starvation-induced differentiation of Dd cells. Moreover, Dd cells have many other advantageous characteristics. Axenic Dd cells are very well suited for conditioned-medium experiments to study the influence of specifically generated Dd EVs upon naive Dd cells, as will be shown in this presentation.

Keywords: exosomes; microvesicles; apoptotic bodies; cancers; *Dictyostelium discoideum*; eukaryotic cell model

Citation: Tatischeff, I. Huge Modification of the Cell Theory by the Recent Discovery of the Widespread Cell-Derived Extracellular Vesicles. *Biol. Life Sci. Forum* **2023**, *21*, 21. https://doi.org/10.3390/blsf2023021021

Academic Editor: Alexander E. Kalyuzhny

Published: 22 March 2023

Funding: This research received no external funding.

Institutional Review Board Statement: Not applicable.

Informed Consent Statement: Not applicable.

Data Availability Statement: Not applicable.

Conflicts of Interest: The author declares no conflict of interest.

biology and life sciences forum

MDPI

Abstract

Electrical Stimulation Causes Activation of c-Src and Focal Adhesion Kinase in Fibroblasts †

Kazuo Katoh

Laboratory of Human Anatomy and Cell Biology, Faculty of Health Sciences, Tsukuba University of Technology, Tsukuba 305-8521, Ibaraki, Japan; katoichi@k.tsukuba-tech.ac.jp

† Presented at Cells, Cells and Nothing but Cells: Discoveries, Challenges and Directions, 6–8 March 2023; Available online: https://cells2023.sciforum.net/.

Abstract: Stimulation of the skin and muscles in the field of acupuncture and moxibustion has been shown to increase local blood flow and metabolism or maintain the body in a sustained healthy state. However, little is known about cellular structural changes in response to the electrical stimulus or how the localization of specific proteins is regulated by such stimulation. Cultured fibroblasts were subject to periodic electrical stimulation for 0 min (the unstimulated control), 2 h, 5 h, and 20 h. After 2 h, cell stress fibers and focal adhesions became enlarged, and the stress fibers exhibited an increase in thickness. Within 20 h of periodic stimulation, both the stress fibers and focal adhesions gradually became larger and thicker. An anti-phosphotyrosine antibody (PY-20) was used to stain the cells after electrical stimulations, which showed the increased staining of focal adhesions. They also exhibited the increased staining of the active form of a focal adhesion kinase (FAK) (pY397) and c-Src (pY418), indicating that electrical stimulation had affected certain proteins associated with signal transduction. An ELISA analysis showed that 20 h of electrical stimulation gradually increased the amount of the active form of c-Src until it was approximately tripled, whereas 5 h of electrical stimulation approximately doubled the amount of the active form of FAK, this being the maximum reached. These findings indicate that electrical stimulation stimulates the activity of c-Src and FAK signaling proteins and alters the structure of the actin cytoskeleton and focal adhesions.

Keywords: electrical stimulation; Src; FAK; focal adhesion kinase; cell adhesion; cytoskeleton; stress fiber; focal adhesion

---

Citation: Katoh, K. Electrical Stimulation Causes Activation of c-Src and Focal Adhesion Kinase in Fibroblasts. *Biol. Life Sci. Forum* **2023**, *21*, 22. https://doi.org/10.3390/blsf2023021022

Academic Editor: Alexander E. Kalyuzhny

Published: 22 March 2023

Funding: The work reported here was supported by Grants-in-Aid for Promotional Projects for Advanced Education and Research, National University Cooperation, Tsukuba University of Technology. The funders had no role in the study design, data collection, analysis, decision to publish, or preparation of the manuscript.

Institutional Review Board Statement: Not applicable.

Informed Consent Statement: Not applicable.

Data Availability Statement: The data presented in this study are available on request from the corresponding author.

Conflicts of Interest: The author declares no conflict of interest.

biology and life sciences forum

MDPI

Abstract

Interleukin (IL)-11 Is Involved in the Functional Liaison between Breast Tumor Cells and the Surrounding Stroma †

Marianna Talia [1,*], Francesca Cirillo [1], Domenica Scordamaglia [1], Maria Francesca Santolla [1], Asia Spinelli [1], Salvatore De Rosis [1], Lucia Muglia [1], Azzurra Zicarelli [1], Anna Maria Miglietta [2], Marcello Maggiolini [1] and Rosamaria Lappano [1]

[1] Department of Pharmacy, Health and Nutritional Sciences, University of Calabria, 87036 Rende, Italy; francesca.cirillo@unical.it (F.C.); domenica.scordamaglia@unical.it (D.S.); mariafrancesca.santolla@unical.it (M.F.S.); asia.spinelli@unical.it (A.S.); salvatore.derosis@libero.it (S.D.R.); lucia.muglia@unical.it (L.M.); azzurra.zicarelli@unical.it (A.Z.); marcello.maggiolini@unical.it (M.M.); rosamaria.lappano@unical.it (R.L.)

[2] Breast Unit, Regional Hospital Cosenza, 87100 Cosenza, Italy; annamariamiglietta@virgilio.it

* Correspondence: marianna.talia@unical.it

† Presented at Cells, Cells and Nothing but Cells: Discoveries, Challenges and Directions, 6–8 March 2023; Available online: https://cells2023.sciforum.net/.

Abstract: Current advances in molecular profiling methodologies and the accessibility of multi-omics datasets are paving the way toward a better understanding of heterogeneous diseases, including breast cancer (BC). In this regard, we sought to uncover the transcriptional changes triggered by estrogen and insulin in a primary BC cell line (BCAHC-1), which expresses the 46kDa isoform of the estrogen receptor (ER)α and the insulin receptor, as we have previously ascertained. Raw data from RNA sequencing of BCAHC-1 cells were processed by the Bcl2Fastq 2.20 version of the Illumina pipeline, while in silico analyses were performed in R Studio using the TCGA dataset. Real-time PCR, immunoblotting, ELISA and chromatin immunoprecipitation experiments were used to identify the molecular events triggered by estrogen and insulin in BCAHC-1 cells and cancer-associated fibroblasts (CAFs). Furthermore, migration and invasion assays allowed us to ascertain the mechanisms triggering these biological responses in the presence of the aforementioned hormone treatments. First, we determined that 17β-estradiol (E2) and insulin stimulate a peculiar IL-11 expression and IL-11 secretion in BCAHC-1 cells. Thereafter, bioinformatics analyses confirmed the up-regulation of IL-11 in ER-positive BCs, with respect to adjacent normal tissues, and its association with worse survival. Next, the involvement of IL-11 in pro-metastatic transduction signaling was established via pathway enrichment analyses. Notably, we found that the secretion of IL-11 by BCAHC-1 cells prompts an invasive phenotype of CAFs through the up-regulation of genes belonging to the extracellular matrix organization pathway, namely, the intercellular adhesion molecule 1 and integrin alpha 5. Overall, our findings indicate that IL-11 secretion by BC cells may elicit a paracrine action on the surrounding stroma and introduce invasive properties, suggesting that IL-11 could be considered a valuable target in comprehensive treatments of ER-positive BC patients.

Keywords: breast cancer; tumor microenvironment; IL-11; bioinformatics

Citation: Talia, M.; Cirillo, F.; Scordamaglia, D.; Santolla, M.F.; Spinelli, A.; De Rosis, S.; Muglia, L.; Zicarelli, A.; Miglietta, A.M.; Maggiolini, M.; et al. Interleukin (IL)-11 Is Involved in the Functional Liaison between Breast Tumor Cells and the Surrounding Stroma. *Biol. Life Sci. Forum* **2023**, *21*, 23. https://doi.org/10.3390/blsf2023021023

Academic Editor: Alexander E. Kalyuzhny

Published: 27 March 2023

Author Contributions: Conceptualization, M.T., F.C., M.M. and R.L.; methodology, M.T., F.C., D.S., M.M. and R.L.; software, M.T. and F.C.; validation, M.T., F.C. and R.L.; formal analysis, M.T.; investigation, M.T., F.C., D.S., M.F.S., A.S., S.D.R., L.M. and A.Z.; resources, A.M.M. and M.M.; data curation, M.T. and F.C.; writing—original draft preparation, M.T. and F.C.; writing—review and editing, M.M. and R.L.; visualization, M.M. and R.L.; supervision, M.M. and R.L.; funding acquisition, M.M. and R.L. All authors have read and agreed to the published version of the manuscript.

Funding: Fondazione AIRC supported M.M. (IG n. 21322) and R.L. (IG n. 27386). Ministero della Salute (Italy) supported M.M. and R.L. (project RF-2019-12368937).

Institutional Review Board Statement: The study was conducted in accordance with the Declaration of Helsinki and approved by the "Comitato Etico Ospedale Regionale, Cosenza, Italy" (approval code: 166, 2 December 2016).

Informed Consent Statement: Written informed consent has been obtained from the patients to publish this paper and the experimental research has been performed with the ethical approval provided by the "Comitato Etico Ospedale Regionale, Cosenza, Italy" (approval code: 166, 2 December 2016).

Data Availability Statement: Not applicable.

Acknowledgments: We would like to thank the special award, namely "Department of Excellence 2018-2022" (Italian Law 232/2016) at the Department of Pharmacy, Health and Nutritional Sciences of the University of Calabria (Italy) and the "Sistema Integrato di Laboratori per l'Ambiente–(SILA) PONa3_00341".

Conflicts of Interest: The authors declare no conflict of interest.

biology and life sciences forum

Abstract

Identification of an Inherent Bioenergetic and Metabolic Phenotype in Late-Onset Alzheimer's Disease †

Kai-Christian Sonntag [1,2,3,*], Woo-In Ryu [1,2,3], Mariana K. Bormann [1,2,3], Eunjung Koh [1,2,3] and Bruce M. Cohen [1,3]

1 Department of Psychiatry, McLean Hospital, Harvard Medical School, Belmont, MA 02478, USA; wryu33014@gmail.com (W.-I.R.); mbormann@mclean.harvard.edu (M.K.B.); ekoh@mclean.harvard.edu (E.K.); bcohen@mclean.harvard.edu (B.M.C.)
2 Basic Neuroscience Division, McLean Hospital, Harvard Medical School, Belmont, MA 02478, USA
3 Program for Neuropsychiatric Research, McLean Hospital, Harvard Medical School, Belmont, MA 02478, USA
* Correspondence: kai@mclean.harvard.edu
† Presented at Cells, Cells and Nothing but Cells: Discoveries, Challenges and Directions, 6–8 March 2023; Available online: https://cells2023.sciforum.net/.

Abstract: The pathology of late-onset Alzheimer's disease (LOAD) is still poorly understood, but it is multifactorial and closely related to changes with aging. We developed a cellular platform for collecting skin fibroblasts or blood cells from LOAD patients and non-demented control individuals, which are then used in an induced pluripotent stem cell (iPSC) paradigm to produce brain cells for determining LOAD pathogenic processes in the context of age, disease, genetic background, cell development, and cell type. This model has provided evidence for an innate inefficient cellular energy management in LOAD that is associated with alterations of cellular transcriptomes and lipid compositions, and interconnected cause-and-effect linkages, such as impaired insulin/IGF-1 signaling, bioenergetic substrate deficiencies, diminished glucose metabolism, and disruption of autophagic flux, among others. In addition, a testing of compounds revealed some restoration of these altered bioenergetic and metabolic processes in LOAD cells. Altogether, our studies have identified an inherent LOAD-associated cellular metabolic phenotype as a potential risk factor for developing neurodegenerative diseases with aging. We propose that our cellular model allows for patient-oriented examination of numerous mechanisms and interactions in LOAD pathogenesis, which can be used as a basis for a personalized medicine approach to predict altered aging and risk of developing dementia, and to test or implement (customized) therapeutic or disease-preventive intervention strategies.

Keywords: autophagy; bioenergetics; brain cells; induced pluripotent stem cells; insulin/IGF-1 signaling; late-onset Alzheimer's disease; metabolism; neurodegeneration; transcriptome

Citation: Sonntag, K.-C.; Ryu, W.-I.; Bormann, M.K.; Koh, E.; Cohen, B.M. Identification of an Inherent Bioenergetic and Metabolic Phenotype in Late-Onset Alzheimer's Disease. *Biol. Life Sci. Forum* **2023**, *21*, 24. https://doi.org/10.3390/blsf2023021024

Academic Editor: Alexander E. Kalyuzhny

Published: 23 March 2023

Author Contributions: Conceptualization, K.-C.S. and B.M.C.; methodology, K.-C.S., B.M.C. and W.-I.R.; validation, K.-C.S., B.M.C. and W.-I.R.; formal analysis, K.-C.S., B.M.C., W.-I.R., M.K.B. and E.K.; investigation, K.-C.S., B.M.C., W.-I.R., M.K.B. and E.K.; resources, K.-C.S. and B.M.C.; data curation, W.-I.R., M.K.B. and E.K.; writing—original draft preparation, K.-C.S.; writing—review and editing, B.M.C.; visualization, K.-C.S.; supervision K.-C.S.; project administration, K.-C.S. and M.K.B.; funding acquisition, B.M.C. All authors have read and agreed to the published version of the manuscript.

Funding: Steele Fund: Program for Neuropsychiatric Research, McLean Hospital (B.M.C.).

Institutional Review Board Statement: The studies were performed in accordance with relevant guidelines and regulations and were approved by the McLean Institutional Review Board (protocol #2007P002597).

Informed Consent Statement: Informed consent was obtained from all subjects involved in the studies.

Data Availability Statement: Data and material are made available upon request.

Acknowledgments: The authors want to thank Suzanne Babb and Fontini Savvides for administrative help.

Conflicts of Interest: The authors declare no conflict of interest.

biology and life sciences forum

MDPI

Abstract

New Insights on Signaling Pathways Deregulated in LAP1-Deficient Cells: A Proteomics Study †

Cátia D. Pereira [1], Guadalupe Espadas [2,3], Filipa Martins [1], Anne T. Bertrand [4], Laurent Servais [5,6], Eduard Sabidó [2,3], Odete A. B. da Cruz e Silva [1] and Sandra Rebelo [1,*]

1 Department of Medical Sciences, Institute of Biomedicine (iBiMED), University of Aveiro, 3810-193 Aveiro, Portugal; daniela.pereira@ua.pt (C.D.P.); samartins@ua.pt (F.M.); odetecs@ua.pt (O.A.B.d.C.e.S.)
2 Centre de Regulació Genòmica (CRG), Barcelona Institute of Science and Technology (BIST), 08003 Barcelona, Spain; guadalupe.espadas@crg.eu (G.E.); eduard.sabido@crg.eu (E.S.)
3 Department of Medicine and Life Sciences, Universitat Pompeu Fabra (UPF), 08003 Barcelona, Spain
4 Institut de Myologie, Center of Research in Myology, INSERM UMRS 974, Medicine Faculty—Sorbonne Université, 75013 Paris, France; a.bertrand@institut-myologie.org
5 MDUK Oxford Neuromuscular Center, Department of Paediatrics, University of Oxford and NIHR Oxford Biomedical Research Center, Oxford OX3 9DU, UK; laurent.servais@paediatrics.ox.ac.uk
6 Neuromuscular Center, Division of Paediatrics, University Hospital of Liège and University of Liège, 4000 Liège, Belgium
* Correspondence: srebelo@ua.pt; Tel.: +351-924-406-306

† Presented at Cells, Cells and Nothing but Cells: Discoveries, Challenges and Directions, 6–8 March 2023; Available online: https://cells2023.sciforum.net/.

Citation: Pereira, C.D.; Espadas, G.; Martins, F.; Bertrand, A.T.; Servais, L.; Sabidó, E.; da Cruz e Silva, O.A.B.; Rebelo, S. New Insights on Signaling Pathways Deregulated in LAP1-Deficient Cells: A Proteomics Study. *Biol. Life Sci. Forum* **2023**, *21*, 25. https://doi.org/10.3390/blsf2023021025

Academic Editor: Alexander E. Kalyuzhny

Published: 24 March 2023

Abstract: Mutations in genes encoding nuclear envelope (NE) proteins, despite being rare, represent a major threat to cell homeostasis by compromising nuclear integrity and function as well as nucleocytoplasmic communication. In the last decade, several diseases have been associated to mutations in the *TOR1AIP1* gene that codes for lamina-associated polypeptide 1 (LAP1), a NE protein ubiquitously expressed in human tissues. Although this is suggestive of an important physiological role of LAP1, it remains unclear which cellular activities are regulated by this protein. To address this, we investigated the molecular repercussions of its deficiency in patient-derived skin fibroblasts carrying a pathological LAP1 mutation (p.E482A), previously reported in a case of severe dystonia, cerebellar atrophy and cardiomyopathy. Using liquid chromatography with tandem mass spectrometry (LC–MS/MS), a quantitative proteome analysis was performed to identify up-/downregulated proteins in LAP1 E482A fibroblasts relative to age-matched control fibroblasts. A subsequent functional characterization of the LC–MS/MS-identified differentially expressed proteins using bioinformatics tools unraveled various signaling pathways/biological processes potentially deregulated in LAP1 E482A fibroblasts, such as DNA repair, neurodevelopment and myogenesis, among others. This work sheds light on dysfunctional molecular mechanisms in LAP1-deficient cells, which will contribute to a better understanding of LAP1's physiological relevance for the maintenance of cell homeostasis and, hopefully, allow the identification of potential therapeutic targets for LAP1-associated pathologies.

Keywords: LAP1; DNA repair; neurodevelopment; myogenesis

Author Contributions: C.D.P., F.M. and S.R. conceived and designed the experiments; C.D.P. and G.E. performed the experimental work and analysis of the results; A.T.B. and L.S. established and provided the human fibroblast cell lines; E.S., O.A.B.d.C.e.S. and S.R. provided the reagents/materials/analysis tools; C.D.P. wrote the original draft of the manuscript; C.D.P., G.E., F.M., E.S., O.A.B.d.C.e.S. and S.R. critically revised the manuscript. All authors have read and agreed to the published version of the manuscript.

Funding: This work was financed by the Institute of Biomedicine (iBiMED)—UIDP/04501/2020 and UIDB/04501/2020—and the Fundação para a Ciência e a Tecnologia (FCT) of the Ministério da

Ciência, Tecnologia e Ensino Superior, the COMPETE 2020 Program, the QREN and the European Union (Fundo Europeu de Desenvolvimento Regional). Authors acknowledge support from EPIC-XS, project number 823839, funded by the Horizon 2020 Program of the European Union. The proteomics analyses were performed in the Proteomics Unit from the Centre de Regulació Genòmica (CRG) and Universitat Pompeu Fabra (UPF). The CRG/UPF Proteomics Unit is part of the Spanish National Infrastructure for Omics Sciences (ICTS OmicsTech). Cátia D. Pereira is the recipient of a PhD fellowship (SFRH/BD/140310/2018 and COVID/BD/152982/2023) co-funded by FCT of the Ministério da Ciência, Tecnologia e Ensino Superior, the Centro 2020 Program and the European Union (Fundo Social Europeu).

Institutional Review Board Statement: Not applicable.

Informed Consent Statement: Not applicable.

Data Availability Statement: The data will be available upon the full length manuscript acceptance.

Acknowledgments: Not applicable.

Conflicts of Interest: The authors declare no conflict of interest.

biology and life sciences forum

MDPI

Abstract

Impaired Nuclear and Mitochondrial Cross-Talk Might Alter mtDNA Epigenetic Regulation in Maternally Inherited Diabetes- and Deafness-Affected Patients †

Luigi Donato [1,2,*], Concetta Scimone [1,2], Simona Alibrandi [1,2,3], Maria Vadalà [2,4], Massimo Castellucci [5], Domenico Mordà [1], Carmela Rinaldi [1], Rosalia D'Angelo [1] and Antonina Sidoti [1]

1 Department of Biomedical and Dental Sciences and Morphofunctional Imaging, Division of Medical Biotechnologies and Preventive Medicine, University of Messina, 98125 Messina, Italy; cscimone@unime.it (C.S.); salibrandi@unime.it (S.A.); domenico.morda@studenti.unime.it (D.M.); crinaldi@unime.it (C.R.); rdangelo@unime.it (R.D.); asidoti@unime.it (A.S.)
2 Department of Biomolecular Strategies, Genetics and Cutting-Edge Therapies, Istituto Euro-Mediterraneo di Scienza e Tecnologia (I.E.ME.S.T.), 90139 Palermo, Italy; maria.vadala@unipa.it
3 Department of Chemical, Biological, Pharmaceutical and Environmental Sciences, University of Messina, 98125 Messina, Italy
4 Department of Biomedicine, Neuroscience and Advanced Diagnostic (BIND), Ophthalmology Institute, University of Palermo, 90143 Palermo, Italy
5 Ophthalmologist Unit, University Hospital Policlinico P. Giaccone, 90127 Palermo, Italy; massimo.castellucci@gmail.com
* Correspondence: ldonato@unime.it; Tel.: +39-0902213136

† Presented at Cells, Cells and Nothing but Cells: Discoveries, Challenges and Directions, 6–8 March 2023; Available online: https://cells2023.sciforum.net/.

Abstract: Mitochondrial pathologies are clinically composite and show highly variable phenotypes amongst all inherited disorders, mainly due to their heteroplasmic nature. Mutations in mitochondrial DNA (mtDNA) and the nuclear genome (gDNA), or both, have been reported in mitochondrial diseases, suggesting common pathophysiological pathways. Nuclear gene mutations identified in mitochondrial diseases are mostly involved in mtDNA replication, transcription and translation, oxidative phosphorylation (OXPHOS), the biosynthesis of mtDNA, nucleoside transport, salvage or synthesis, and the homeostasis of mitochondrial deoxyribonucleoside triphosphates (dNTP) pool. The m.3243 A>G mtDNA mutation in the MT-TL1 gene coding for the tRNALeu (UUR) is one of the most common mitochondrial disease-causing mutations, with a carrier rate as high as 1:400. Recent studies suggest that patients with the m.3243 A>G mutation present a huge clinical heterogeneity supporting the necessity to investigate the nuclear genome to improve the knowledge on composite mitochondrial disorders, such as mitochondrial encephalomyopathy, lactic acidosis and stroke-like episodes (MELAS), maternally inherited diabetes and deafness (MIDD) and myopathy. MIDD is a multi-system disorder characterized by diabetes, hearing impairment, and maculopathy but can present several other clinical manifestations. The present study aimed to analyze the whole mitochondrial genome and the whole exome of a clinically characterized MIDD family, negative to the m.3243 A>G variant, and identify mutations in both gDNA and mtDNA, as well as their biological role in their heterogeneous phenotype. The obtained results permitted us to hypothesize that the mitochondrial defects might be due to the epigenetic deregulation of the mitochondrial and nuclear-encoded genes coding for mitochondrial structure and functions. Thus, epigenetic modifications in the context of mitochondrial dysfunctions represent an emerging area of research, possibly useful for innovative mtDNA-related disease differential analyses.

Keywords: mtDNA; WGS; WES; epigenetics; MIDD

Citation: Donato, L.; Scimone, C.; Alibrandi, S.; Vadalà, M.; Castellucci, M.; Mordà, D.; Rinaldi, C.; D'Angelo, R.; Sidoti, A. Impaired Nuclear and Mitochondrial Cross-Talk Might Alter mtDNA Epigenetic Regulation in Maternally Inherited Diabetes- and Deafness-Affected Patients. *Biol. Life Sci. Forum* **2023**, *21*, 26. https://doi.org/10.3390/blsf2023021026

Academic Editor: Alexander E. Kalyuzhny

Published: 24 March 2023

Author Contributions: Conceptualization, L.D.; methodology, L.D., M.V. and M.C.; software, L.D. and C.S.; validation, C.S., S.A. and D.M.; formal analysis, L.D.; investigation, L.D. and D.M.; resources, C.S.; data curation, L.D. and C.S.; writing—original draft preparation, L.D.; writing—review and editing, C.S., R.D. and C.R.; visualization, C.R. and R.D.; supervision, A.S.; project administration, A.S. All authors have read and agreed to the published version of the manuscript.

Funding: This research received no external funding.

Institutional Review Board Statement: The study was conducted in accordance with the Declaration of Helsinki, and approved by Ethics Committee of University of Messina (protocol code 23/17bis).

Informed Consent Statement: Informed consent was obtained from all subjects involved in the study.

Data Availability Statement: Data is contained within the article.

Conflicts of Interest: The authors declare no conflict of interest.

biology and life sciences forum

MDPI

Abstract

Modulation of Retinoic Acid Receptor Signaling Pathway via All-Trans Retinoic Acid in Merkel Cell Carcinoma Cells †

John Charles Rotondo [1,*], Chiara Mazziotta [1], Giampaolo Morciano [1], Christian Felice Cervellera [1], Giada Badiale [1], Giulia Tonnini [1], Giulia Di Mauro [1], Paolo Pinton [1], Antoine Touzé [2], Pauline Gaboriaud [2], Mauro Tognon [1] and Fernanda Martini [1]

1 Department of Medical Sciences, University of Ferrara, 44121 Ferrara, Italy; mzzchr@unife.it (C.M.); mrcgpl@unife.it (G.M.); christianfelice.cervellera@unife.it (C.F.C.); giada.badiale@unife.it (G.B.); giulia.tonnini@edu.unife.it (G.T.); giulia.dimauro@unife.it (G.D.M.); paolo.pinton@unife.it (P.P.); mauro.tognon@unife.it (M.T.); fernanda.martini@unife.it (F.M.)
2 ISP "Biologie des infections à polyomavirus" Team, UMR INRAE 1282, University of Tours, 37200 Tours, France; antoine.touze@univ-tours.fr (A.T.); pauline.gaboriaud@univ-tours.fr (P.G.)
* Correspondence: rtnjnc@unife.it; Tel.: +39-0532-455-536

† Presented at Cells, Cells and Nothing but Cells: Discoveries, Challenges and Directions, 6–8 March 2023; Available online: https://cells2023.sciforum.net/.

Abstract: The biological activity of retinoic acid or all-trans retinoic acid (ATRA) is mediated by retinoic receptors, which are ligand-dependent transcription factors that activate genes crucial for cell differentiation. Dysregulations of retinoic receptor signaling pathway led to carcinogenesis. A strong in vitro/in vivo antitumor activity of ATRA by modulating the retinoic pathway has been proved in carcinoma of different histotypes. However, the effect of this molecule in Merkel cell carcinoma (MCC), a rare but aggressive skin neoplasm of viral origin in 80% of cases, is unknown. Herein, we investigated the antineoplastic effect of ATRA in Merkel cell polyomavirus (MCPyV)-positive/-negative MCC cells and in human fibroblasts as controls. The antineoplastic effect of ATRA was evaluated at day 3 of treatment by testing MCC cell proliferation, migration, and clonogenicity. Apoptosis/cell death and cell cycle were evaluated via Annexin-V/propidium iodide (P.I.) and TALI assays, respectively. Apoptotic and retinoic pathways were evaluated by RT^2 Profiler PCR mRNA array, which allows the analysis of pro/anti-apoptotic and retinoic pathway genes (84 + 84 genes), as well as by Western blot (WB) analysis. ATRA treatment led to a strong reduction in MCC cell proliferation, migration and clonogenicity, while inducing cell cycle arrest and promoting apoptosis/death in MCC cells, with a more pronounced effect in MCPyV-positive MCC cells. A significant overexpression of various pro-apoptotic markers in ATRA-treated MCC cells compared to untreated cells was determined by gene expression array and WB analyses. No phenotypic and molecular effects were identified in ATRA-treated fibroblast control cells. Upon ATRA treatments in MCC cells, numerous retinoic signaling genes, such as *BMP2*, *FOXA1*, *MAFB*, *RBP4*, *OLIG2*, *UCP1* were found to be differentially expressed compared to untreated cells. Our in vitro data indicate that ATRA is effective in reducing MCC cell growth while presenting strong pro-apoptotic effects and favoring cell cycle arrest/death via retinoic receptor signaling pathway regulation.

Keywords: Merkel cell carcinoma; Merkel cell polyomavirus; all-trans retinoic acid

Citation: Rotondo, J.C.; Mazziotta, C.; Morciano, G.; Cervellera, C.F.; Badiale, G.; Tonnini, G.; Di Mauro, G.; Pinton, P.; Touzé, A.; Gaboriaud, P.; et al. Modulation of Retinoic Acid Receptor Signaling Pathway via All-Trans Retinoic Acid in Merkel Cell Carcinoma Cells. *Biol. Life Sci. Forum* **2023**, *21*, 27. https://doi.org/10.3390/blsf2023021027

Academic Editor: Alexander E. Kalyuzhny

Published: 24 March 2023

Author Contributions: Conceptualization, J.C.R. and F.M.; methodology, C.M.; software, G.M.; validation, C.F.C., G.B. and G.T.; formal analysis, C.M.; investigation, G.D.M.; resources, P.P., A.T., P.G.; data curation, C.M.; writing—original draft preparation, J.C.R.; writing—review and editing, J.C.R.; supervision, A.T. and M.T.; project administration, F.M. All authors have read and agreed to the published version of the manuscript.

Funding: This work was supported, in part, by grants MFAG 21956 (to J.C.R.), IG 21617 (to M.T.) from the Associazione Italiana per la Ricerca sul Cancro (AIRC), Milan, Italy, and University of Ferrara,

FAR grants 2021 (F.M. and M.T.). J.C.R. was supported by Umberto Veronesi Foundation. C.M. was supported by a AIRC fellowship for Italy (ID: 26829) and by "Bando Giovani anno 2022 per progetti di ricerca finanziati con il contributo 5 × 1000 anno 2020".

Institutional Review Board Statement: Not applicable.

Informed Consent Statement: Not applicable.

Data Availability Statement: Not applicable.

Conflicts of Interest: The authors declare no conflict of interest.

biology and life sciences forum

MDPI

Abstract

Maternal Hyperhomocysteinemia Disturbs the Brain Development and Maturation in Offspring †

Dmitrii S. Vasilev [1,*], Anastasiia D. Shcherbitskaia [1,2], Natalia L. Tumanova [1], Yulia P. Milyutina [2], Anna A. Kovalenko [1], Nadezhda M. Dubrovskaya [1], Anastasiia V. Mikhel [1,2], Daria B. Inozemtseva [2], Irina V. Zalozniaia [2] and Alexander V. Arutjunyan [2]

1 I.M. Sechenov Institute of Evolutionary Physiology and Biochemistry of the Russian Academy of Sciences, St. Petersburg 194223, Russia; nastusiq@gmail.com (A.D.S.); tuman-1946@mail.ru (N.L.T.); kovalenko_0911@mail.ru (A.A.K.); ndub@mail.ru (N.M.D.); anastasia.michel39@gmail.com (A.V.M.)

2 D.O. Ott Institute of Obstetrics, Gynecology and Reproductive Medicine, St. Petersburg 199034, Russia; milyutina1010@mail.ru (Y.P.M.); inozemtseva400@gmail.com (D.B.I.); irinabiolog@rambler.ru (I.V.Z.); alexarutiunjan@gmail.com (A.V.A.)

* Correspondence: dvasilyev@bk.ru

† Presented at Cells, Cells and Nothing but Cells: Discoveries, Challenges and Directions, 6–8 March 2023; Available online: https://cells2023.sciforum.net/.

Abstract: The effect of the homocysteine toxicity on both mother and embryo is known to induce disruption of placental blood flow and disturbances of the brain formation in offspring. The mechanisms of these effects are poorly understood and should be studied. The effects of prenatal hyperhomocysteinemia (pHHC) on the expression of some neuronal genes, neural tissue maturation and neuronal migration were analyzed in this study. Hyperhomocysteinemia was induced in female rats by per os administration of 0.15% aqueous methionine solution during pregnancy. On P5–P20 some features of developmental delay were observed in both cortical and hippocampus tissue ultrastructure in pHHC pups, accompanied by a retardation in body weight and motor development. In hippocampus tissue of P20 pHHC pups of synaptic glomeruli were absence suggesting more essential tissue immaturity compared to the cortical one. In pHHC pupst was shown decreased number and disturbed positioning of the neuronal cells labeled on E14 or E18, suggesting decrease in generation of cortical neuroblasts and disturbance in their radial migration into the cortical plate. On E14 the expression of the *Kdr* gene (an angiogenesis system component) was decreased in pHHC fetus brains. The content of SEMA3E and the MMP-2 activity level was increased. On E20 the increase in proBDNF/mBDNF ratio was also shown in pHHC pups, it might affect positioning maturation and viability of neuronal cell. The activation of caspase-3 accompanied by decrease in the level of procaspase-8 in the brain tissue of E20 pHHC fetuses may suggest the presence of cell apoptosis. It can be concluded that pHHC disturbs the mechanisms of early brain development and delay in brain tissue maturation in both neocortex and hippocampus of pups during early postnatal ontogenesis.

Keywords: hyperhomocysteinemia; rat; hippocampus; brain cortex

Citation: Vasilev, D.S.; Shcherbitskaia, A.D.; Tumanova, N.L.; Milyutina, Y.P.; Kovalenko, A.A.; Dubrovskaya, N.M.; Mikhel, A.V.; Inozemtseva, D.B.; Zalozniaia, I.V.; Arutjunyan, A.V. Maternal Hyperhomocysteinemia Disturbs the Brain Development and Maturation in Offspring. *Biol. Life Sci. Forum* **2023**, *21*, 28. https://doi.org/10.3390/blsf2023021028

Academic Editor: Alexander E. Kalyuzhny

Published: 27 March 2023

Author Contributions: Conceptualization, A.D.S. and D.S.V.; methodology and data analysis, A.D.S., D.S.V., N.L.T., A.A.K., N.M.D., Y.P.M., I.V.Z., D.B.I. and A.V.M.; writing—original draft preparation, A.D.S. and D.S.V.; writing—review and editing, N.L.T., N.M.D. and A.V.A.; supervision and funding acquisition, D.S.V. and A.V.A. All authors have read and agreed to the published version of the manuscript.

Funding: Supported by the RFBR #20 015-00388, Russian state budget assignment for IEPhB (#075-00408-21-00) and D.O. Ott Research Institute of Obstetrics, Gynecology and Reproductology (#1021062812133-0-3.2.2).

Institutional Review Board Statement: The study was conducted according to the guidelines of the Declaration of Helsinki, and approved by the Institutional Ethics Committees of I.M. Sechenov Institute of Evolutionary Physiology and Biochemistry RAS (protocol code 3/2020 as of 18 March 2020).

Informed Consent Statement: Not applicable.

Data Availability Statement: The data presented in this study are available on request from the corresponding author.

Conflicts of Interest: The authors declare no conflict of interest.

biology and life sciences forum

MDPI

Abstract

The Rab11 Family Controls Signalling to the Cytoskeleton for Cell Migration and Invasion †

Patrick Caswell *, Jakub Gemperle, Chloe Flett and Beverley Wilson

Wellcome Trust Centre for Cell-Matrix Research, Faculty of Biology Medicine and Health, University of Manchester, Oxford Road, Manchester M13 9PT, UK; jakub.gemperle@gmail.com (J.G.); chloe.flett@manchester.ac.uk (C.F.); beverley.wilson27@gmail.com (B.W.)

* Correspondence: patrick.caswell@manchester.ac.uk

† Presented at Cells, Cells and Nothing but Cells: Discoveries, Challenges and Directions, 6–8 March 2023; Available online: https://cells2023.sciforum.net/.

Abstract: Endocytic recycling controls the return of internalised cargos to the plasma membrane to coordinate their positioning, availability and downstream signalling. The Rab4 and Rab11 small GTPase families regulate distinct recycling routes, broadly classified as fast recycling from early endosomes (Rab4) and slow recycling from perinuclear recycling endosomes (Rab11), and both routes handle a broad range of overlapping cargos to regulate cell behaviour. We have previously shown that Rab11 regulates the recycling of integrins to promote cancer cell migration and invasion, at least in part by controlling RhoGTPase activity at the leading edge, but the mechanisms that underpin cytoskeleton regulation by Rab11 family members are still unclear. We adopted a proximity labelling approach, BioID, to identify and compare the protein complexes recruited by Rab4a, Rab11a and Rab25 (a Rab11 family member implicated in cancer aggressiveness), revealing robust protein–protein interaction networks of well-characterised, new cargos and trafficking machinery in migratory cancer cells. Gene ontological analysis of these interconnected networks revealed that these endocytic recycling pathways are intrinsically connected to cell motility and cell adhesion, and we demonstrate that several of these new Rab11 and Rab25 associated proteins are required for efficient cancer cell migration in a 3D matrix. Rab11 and Rab25 vesicles are found at the perinuclear recycling compartment but also in the tips of protrusions in cells moving in a 3D matrix. This leads us to speculate that these recycling pathways deliver cargos to directly promote protrusion formation and extension. To test this, we established a magneto-genetic approach to physically re-localise Rab25 vesicles in cells in 2D and 3D matrices. Using this technique, we are able to show that repositioning of Rab25 vesicles to the cell cortex promotes the formation of protrusions in a manner dependent on actin-polymerising protein formins, but not Arp2/3. Together, these data reveal a direct role for Rab11 family members in directing cytoskeletal signalling to promote cancer cell invasion.

Keywords: cell migration; cytoskeleton; endocytic recycling; Rab11; formins

Citation: Caswell, P.; Gemperle, J.; Flett, C.; Wilson, B. The Rab11 Family Controls Signalling to the Cytoskeleton for Cell Migration and Invasion. *Biol. Life Sci. Forum* **2023**, *21*, 29. https://doi.org/10.3390/blsf2023021029

Academic Editor: Alexander E. Kalyuzhny

Published: 27 March 2023

Author Contributions: Conceptualization, P.C., J.G. and B.W.; methodology, P.C., J.G., C.F. and B.W.; investigation, P.C., J.G., C.F. and B.W., funding acquisition, P.C. and J.G. All authors have read and agreed to the published version of the manuscript.

Funding: Cancer Research UK (DCRPGF\100002) and European Union's Horizon 2020 research and innovation programme under grant agreement No [836212].

Institutional Review Board Statement: Not applicable.

Informed Consent Statement: Not applicable.

Data Availability Statement: Not applicable.

Conflicts of Interest: The authors declare no conflict of interest.

biology and life sciences forum

MDPI

Abstract

Regulation of Extrasynaptic Glutamatergic Signaling by Polysialylated NCAM in Health and Disease †

Alexander Dityatev [1,2,3]

1 Molecular Neuroplasticity Group, German Center for Neurodegenerative Diseases (DZNE), 39120 Magdeburg, Germany; alexander.dityatev@dzne.de
2 Medical Faculty, Otto-von-Guericke-University of Magdeburg, 39120 Magdeburg, Germany
3 Center for Behavioral Brain Sciences (CBBS), 39106 Magdeburg, Germany

† Presented at Cells, Cells and Nothing but Cells: Discoveries, Challenges and Directions, 6–8 March 2023; Available online: https://cells2023.sciforum.net/.

Abstract: The neural cell adhesion molecule NCAM is known to mediate cell–to–cell and cell–to–extracellular matrix (ECM) adhesion via homophilic and heterophilic interactions. During brain development, NCAM and the associated glycan, polysialic acid (polySia), play important roles in cell migration proliferation, neurite outgrowth and fasciculation, and synaptogenesis. In the adult rodent brain, NCAM regulates synaptic plasticity, learning, and memory. Dysregulated cortical expression of NCAM and polySia has been reported in Alzheimer's disease and schizophrenia. Our data demonstrate i) the importance of polySia–NCAM in the balancing of signaling through synaptic/extrasynaptic NMDA receptors and ii) the therapeutic value of short defined-length polySia fragments to restrain GluN2B-mediated signaling in several animal models of neurological and psychiatric diseases.

Keywords: LTP; synaptic plasticity; polysialic acid; schizophrenia; Alzheimer's disease; NMDA receptor

Citation: Dityatev, A. Regulation of Extrasynaptic Glutamatergic Signaling by Polysialylated NCAM in Health and Disease. *Biol. Life Sci. Forum* **2023**, *21*, 30. https://doi.org/10.3390/blsf2023021030

Academic Editor: Alexander E. Kalyuzhny

Published: 3 April 2023

Funding: This work has been supported by the Italian Institute of Technology and DZNE, the European Union's Horizon 2020 Research and Innovation Programme under the Marie Sklodowska-Curie grant agreement no. 642881 (ECMED), and DFG SFB 1436 TP A03 (Project-ID 425899996).

Institutional Review Board Statement: All treatments and behavioral procedures were conducted in accordance with animal research ethics standards defined by German law and approved by the Ethical Committee on Animal Health and Care of the State of Saxony-Anhalt (42502-2-1159 DZNE, 42502-2-1343 DZNE, and 42502-2-1346 DZNE) or by the Italian Committee on Animal Health and Care of the governmental body in Rome.

Informed Consent Statement: Not applicable.

Data Availability Statement: Data will be made available on request.

Acknowledgments: I thank Jenny Schneeberg and Katrin Boehm for excellent technical assistance, Inseon Song for helpful discussions of patch-clamp recordings, Rahul Kaushik for helpful discussions of immunohistochemistry, Maura Ferrer for help with spine imaging, Pierre Paoletti for the GluN1 expression vector, and Hristo Varbanov, Shaobo Jia, Gaga Kochlamazashvili, Subhrajit Bhattacharya, Manal Ali Buabeid, Mohamed El Tabbal, Hussam Hayani, Stoyan Stoyanov, Weilun Sun, Oleg Senkov, Hauke Thiesler, Iris Röckle, Herbert Hildebrandt, Eckart Gundelfinger, Vishnu Suppiramaniam, and Rita Gerardy-Schahn for their contributions and support in analysis of polySia functions.

Conflicts of Interest: A.D. filed an international patent application on "Polysialic acid and derivatives thereof, pharmaceutical composition and method of producing polysialic acid", WO2020025653A2. A.D. is the editor-in-chief in *Cell microenvironment* section of *Cells*.

biology and life sciences forum

MDPI

Abstract

Probiotic Engineering: Resolving How Fermentable Sugars Affect Aggregation, Adhesion, and Aggression in Lactobacillaceae †

Ronit Suissa [1], Michael Meijler [1], Omry Koren [2] and Ilana Kolodkin-Gal [3,*]

1 Department of Chemistry, The National Institute for Biotechnology in the Negev, Ben-Gurion University of the Negev, Beer-Sheva 8410501, Israel; ronit.suissa@post.bgu.ac.il (R.S.); meijler@bgu.ac.il (M.M.)
2 Azrieli Faculty of Medicine, Bar-Ilan University, Safed 1311502, Israel; omry.koren@biu.ac.il
3 Department of Plant Pathology and Microbiology, The Robert H. Smith Faculty of Agriculture, Food and Environment, The Hebrew University of Jerusalem, Rehovot 7610001, Israel
* Correspondence: ilana.kolodkin-gal@weizmann.ac.il

† Presented at Cells, Cells and Nothing but Cells: Discoveries, Challenges and Directions, 6–8 March 2023; Available online: https://cells2023.sciforum.net/.

Abstract: Lactobacillaceae are Gram-positive and lactic acid-positive (LAB) bacteria that frequently serve as probiotics. LAB strains vary in their responses to different carbohydrates as free-living and biofilm communities. We previously found that fermentable sugars triggered an altered carrying capacity with strain specificity during planktonic growth, calling for adding a buffering system during the formulation of probiotics. In addition, a heterogeneous response to fermentable sugars was manifested in microbial aggregation (measured by image-stream flow cytometry), colony development, and attachment to mucin. Of all the probiotic strains, *L. rhamnosus GG* (LGG), a prevalent probiotic species, manifested an enhanced survival of self-imposed acid stress, consistent with the enhanced cell wall modulation observed by transmitting electron microscopy and proteomic analysis. A comprehensive proteomic and metabolomic study revealed that the formation of biofilms and aggregation capacity is a specific response to glucose independent of self-imposed acid stress. In contrast, the increased competitiveness and aggression of LGG and other LAB strains towards enteric pathogens were a synergistic outcome of a change in organic acid production, glucose-dependent bacteriocin production, and fermentation-specific volatile production. Our improved resolution into the cellular circuits (metabolome, proteome, and volatilome) of probiotic strains and their interactions can lead to developing novel therapeutic approaches to combat GI tract infections.

Keywords: probiotics; microbiome; metabolome; bacteriocins

Citation: Suissa, R.; Meijler, M.; Koren, O.; Kolodkin-Gal, I. Probiotic Engineering: Resolving How Fermentable Sugars Affect Aggregation, Adhesion, and Aggression in Lactobacillaceae. *Biol. Life Sci. Forum* **2023**, *21*, 31. https://doi.org/10.3390/blsf2023021031

Academic Editor: Alexander E. Kalyuzhny

Published: 4 April 2023

Author Contributions: Conceptualization, R.S. and I.K.-G.; resources, M.M., I.K.-G. and O.K.; data curation, I.K.-G.; writing—original draft preparation, I.K.-G.; writing—review and editing, I.K.-G.; visualization, I.K.-G.; supervision, M.M., I.K.-G. and O.K.; project administration, R.S., M.M. and I.K.-G.; funding acquisition, I.K.-G. All authors have read and agreed to the published version of the manuscript.

Funding: This research was funded by ImoH grant number 713645 to I.K.-G.

Institutional Review Board Statement: Not applicable.

Informed Consent Statement: Not applicable.

Data Availability Statement: Not applicable.

Conflicts of Interest: The authors declare no conflict of interest.

biology and life sciences forum

MDPI

Proceeding Paper

Exploring the Effect of PAK Inhibition in a 3D Pancreatic Cancer Invasion Model †

Marianne Best [1,*], Debashis Sarker [2] and Claire M. Wells [1]

1 School of Cancer and Pharmaceutical Sciences, Faculty of Life Sciences and Medicine, King's College London, London SE1 1UL, UK; claire.wells@kcl.ac.uk
2 Guy's and St Thomas' NHS Foundation Trust, London SE1 9RT, UK; debashsis.sarker@kcl.ac.uk
* Correspondence: marianne.best@kcl.ac.uk

† Presented at Cells, Cells and Nothing but Cells: Discoveries, Challenges and Directions, 6–8 March 2023; Available online: https://cells2023.sciforum.net/.

Abstract: Pancreatic Ductal Adenocarcinoma (PDAC) is an aggressive cancer, with over half of patients presenting with metastatic PDAC at diagnosis. Most patients receive conventional chemotherapy which invariably faces resistance, and a key facilitator in this is the PDAC stroma which acts as a functional mediator of disease progression through bilateral crosstalk between stromal cells and cancer cells. 'Migrastatics' are a new drug class which target cell migration pathway effector proteins to attenuate cancer cell invasion. Improvement in PDAC treatment strategy is well-overdue and migrastatics as adjuvant therapy is one avenue gaining traction. The p21-activated kinase (PAK) family is frequently overexpressed and/or amplified in PDAC where it regulates cytoskeletal actin contractility as well as transcription. Pre-clinical PAK inhibitors have shown reduced PDAC cell invasion in vitro, yet it is unknown how the PDAC stromal cells would respond to a PAK inhibitor and how this could consequently affect PDAC invasion. My PhD project investigates the Pancreatic stellate cell response to PAK inhibition.

Keywords: pancreatic cancer; cell migration; cell invasion; p21-activated kinases (PAKs); kinase inhibitors; actomyosin contractility; cytoskeletal remodeling; transcription; migrastatics; 3D models

Citation: Best, M.; Sarker, D.; Wells, C.M. Exploring the Effect of PAK Inhibition in a 3D Pancreatic Cancer Invasion Model. *Biol. Life Sci. Forum* **2023**, *21*, 32. https://doi.org/10.3390/blsf2023021032

Academic Editor: Alexander E. Kalyuzhny

Published: 14 April 2023

1. Introduction

Pancreatic ductal adenocarcinoma (PDAC) is an aggressive and rapidly invasive cancer, with only 10% of patients surviving 5 years post-diagnosis. Chemotherapy treatment invariably faces resistance, and a central facilitator of this is the PDAC stroma, which acts as a functional mediator of disease progression through bilateral crosstalk between PDAC cells and stromal cells [1]. The p21-activated kinases (PAK1–6) regulate cytoskeletal actin dynamics as well as cellular transcription, and are frequently overexpressed and/or amplified in PDAC to promote cancer cell migration [2].

Cancer Research UK is developing PAK inhibitors to be antimigration cancer therapeutics called 'migrastatics'. Thus far, preclinical PAK inhibitors have shown promising results by attenuating the 3D invasion of PDAC cells in vitro [3], yet the stromal response to PAK inhibition remains unknown. Pancreatic stellate cells (PSCs) are key stromal players in PDAC, and it has been shown that drug administration can alter PSCs' behavior to ultimately drive overall therapeutic outcomes [4,5]. Therefore, my PhD project investigates the PSCs' responses to PAK inhibition with regard to 3D PDAC invasion.

2. Methods

A 3D spheroid assay is used to co-culture PDAC cells and PSCs together to model PDAC invasion, and subsequently investigate the effect of a PAK inhibitor. Immunofluorescence, western blotting, and gel contraction assays are used to characterize PSC behavior and explore PAK expression. A novel pipeline was developed to isolate PDAC cells and

PSCs from 3D co-culture spheroids for downstream RNA-sequencing. All sequencing analysis is performed using R.

3. Results

Characterization studies compared immortalized stellate cell model, PS-1 against the commercially available HPaSteC, validating that the latter was the more representative model to bring forward. Exploration of PSC PAK expression was completed in both Pancreatic Stellate cell models to observe how group specific expression may differ.

Optimization of the 3D co-culture spheroid was completed with a panel of PDAC cell lines, choosing PATU8902 PDAC cells to take forward in the invasion model. We next investigated the effect of pan-PAK inhibitor treatment in the 3D PDAC: Stellate co-culture setting to show invasion is reduced, and current work is investigating this further.

In addition to cytoskeletal dynamics, PAKs have strong links to transcriptional regulation. We developed a pipeline to isolate PDAC and PSCs from embedded 3D invaded spheroids for downstream RNA-sequencing in order to evaluate the transcriptomic landscape of both PDAC and PSC compartments under PAK inhibition. So far, quality control shows good quality RNA was obtained and we were able to isolate out the two cell types, and differential gene expression will be explored next.

4. Conclusions

Our findings suggest that PSCs are an important cell type in promoting PDAC invasion and should be considered in therapeutic development. Current work is investigating how PAK inhibition could affect PSC-driven PDAC invasion. RNA-sequencing analysis is underway to explore the differentially expressed genes in PAK-inhibited PDAC and stellate cells, and to understand how PAK inhibition may influence the crosstalk between these two cell types.

Author Contributions: Conceptualization, C.M.W. and D.S.; investigation, M.B.; writing, M.B.; supervision, C.M.W. and D.S. All authors have read and agreed to the published version of the manuscript.

Funding: This research was funded by Cancer Research UK.

Institutional Review Board Statement: Not applicable.

Informed Consent Statement: Not applicable.

Data Availability Statement: Not applicable.

Conflicts of Interest: The authors declare no conflict of interest.

References

1. Biffi, G.; Oni, T.E.; Spielman, B.; Hao, Y.; Elyada, E.; Park, Y.; Preall, J.; Tuveson, D.A. IL1-induced Jak/STAT signaling is antagonized by TGFβ to shape CAF heterogeneity in pancreatic ductal adenocarcinoma. *Cancer Discov.* **2019**, *9*, 282–301. [CrossRef] [PubMed]
2. King, H.; Nicholas, N.S.; Wells, C.M. Role of p-21-activated kinases in cancer progression. *Int. Rev. Cell Mol. Biol.* **2014**, *309*, 347–387. [CrossRef] [PubMed]
3. King, H.; Thillai, K.; Whale, A.; Arumugam, P.; Eldaly, H.; Kocher, H.M.; Wells, C.M. PAK4 interacts with p85 alpha: Implications for pancreatic cancer cell migration. *Sci. Rep.* **2017**, *7*, 1–12. [CrossRef] [PubMed]
4. Cukierman, E. The Few yet Fab p4 ulous Pancreatic Stellate Cells Give Rise to Protumoral CAFs. *Cancer Discov.* **2022**, *12*, 296–298. [CrossRef] [PubMed]
5. Öhlund, D.; Handly-Santana, A.; Biffi, G.; Elyada, E.; Almeida, A.S.; Ponz-Sarvise, M.; Corbo, V.; Oni, T.E.; Hearn, S.A.; Lee, E.J. Distinct populations of inflammatory fibroblasts and myofibroblasts in pancreatic cancer. *J. Exp. Med.* **2017**, *214*, 579–596. [CrossRef] [PubMed]

biology and life sciences forum

MDPI

Abstract

Human Pluripotent Stem Cells from Diabetic and Nondiabetics Improve Retinal Pathology in Diabetic Mice †

Chang-Hyun Gil [1,2], Dibyendu Chakraborty [3], Cristiano P. Vieira [3], Nutan Prasain [1,4], Sergio Li Calzi [3], Seth D. Fortmann [3,5], Ping Hu [3], Kimihiko Banno [1,6], Mohamed Jamal [7,8], Chao Huang [3], Micheli S. Sielski [3], Yang Lin [1,9], Xinxin Huang [10,11], Mariana D. Dupont [3], Jason L. Floyd [3], Ram Prasad [3], Ana Leda F. Longhini [3,12], Trevor J. McGill [13], Hyung-Min Chung [14], Michael P. Murphy [2], Darrell N. Kotton [7], Michael E. Boulton [3], Mervin C. Yoder [1,15] and Maria B. Grant [3,*]

1 Department of Pediatrics, Indiana University School of Medicine, Indianapolis, IN 46202, USA; chgil@iu.edu (C.-H.G.); nprasain@gmail.com (N.P.); kbanno@naramed-u.ac.jp (K.B.); yal4002@med.cornell.edu (Y.L.); myoder@iu.edu (M.C.Y.)
2 Department of Surgery, Indiana University School of Medicine, Indianapolis, IN 46202, USA; mipmurph@iupui.edu
3 Department of Ophthalmology and Visual Sciences, University of Alabama at Birmingham (UAB), Birmingham, AL 35294, USA; dchakrab@uab.edu (D.C.); cvieira@uabmc.edu (C.P.V.); scalzi@uabmc.edu (S.L.C.); sfortma2@uab.edu (S.D.F.); huping3366@gmail.com (P.H.); huangchao747@gmail.com (C.H.); micheli.sielski@gmail.com (M.S.S.); mdupont@uab.edu (M.D.D.); jfloyd1@uab.edu (J.L.F.); ramprasad@uabmc.edu (R.P.); allonghini@gmail.com (A.L.F.L.); meboulton@uabmc.edu (M.E.B.)
4 Astellas Institute for Regenerative Medicine (AIRM), Westborough, MA 01581, USA
5 Medical Scientist Training Program (MSTP), School of Medicine, University of Alabama at Birmingham, Birmingham, AL 35294, USA
6 Department of Physiology II, Nara Medical University, Kashihara, Nara 634-8521, Japan
7 Center for Regenerative Medicine, Pulmonary Center, Department of Medicine, Boston University School of Medicine, Boston, MA 02118, USA; mohamed.jamal@mbru.ac.ae (M.J.); dkotton@bu.edu (D.N.K.)
8 Department of Endodontics, Hamdan Bin Mohammed College of Dental Medicine, Mohammed Bin Rashid University of Medicine and Health Sciences, Dubai P.O. Box 505055, United Arab Emirates
9 Department of Medicine, Ansary Stem Cell Institute, Weill Cornell Medicine, New York, NY 10021, USA
10 Department of Microbiology and Immunology, Indiana University School of Medicine, Indianapolis, IN 46202, USA; xinxinhuang@fudan.edu.cn
11 Zhongshan-Xuhui Hospital and Shanghai Key Laboratory of Medical Epigenetics, Institutes of Biomedical Sciences, Fudan University, Shanghai 310104, China
12 Flow Cytometry Core Facility, Memorial Sloan Kettering Cancer Center (MSKCC), New York, NY 10065, USA
13 Department of Ophthalmology, Casey Eye Institute, Oregon Health and Science University, Portland, OR 97239, USA; mcgilltr@ohsu.edu
14 Department of Stem Cell Biology, School of Medicine, Konkuk University, Seoul 05029, Republic of Korea; hmchung@kku.ac.kr
15 Indiana Center for Regenerative Medicine and Engineering, Indiana University School of Medicine, Indianapolis, IN 46202, USA
* Correspondence: mariagrant@uabmc.edu
† Presented at Cells, Cells and Nothing but Cells: Discoveries, Challenges and Directions, 6–8 March 2023; Available online: https://cells2023.sciforum.net/.

Citation: Gil, C.-H.; Chakraborty, D.; Vieira, C.P.; Prasain, N.; Calzi, S.L.; Fortmann, S.D.; Hu, P.; Banno, K.; Jamal, M.; Huang, C.; et al. Human Pluripotent Stem Cells from Diabetic and Nondiabetics Improve Retinal Pathology in Diabetic Mice. *Biol. Life Sci. Forum* **2023**, *21*, 33. https://doi.org/10.3390/blsf2023021033

Academic Editor: Alexander E. Kalyuzhny

Published: 14 April 2023

Keywords: human-induced pluripotent stem cells; vascular repair; diabetes; diabetic retinopathy

Human-induced pluripotent stem cells (hiPSCs) cells have the proliferative potential and ability to differentiate into numerous cell types. We have previously generated vascular wall-derived reparative cells called endothelial colony-forming cells (ECFCs) from iPSCs derived from well-characterized, healthy, and diabetic individuals [1]. Our studies showed that these human iPSCs incorporate into blood vessels when implanted subcutaneously into immune compromised mice. These cells, of either diabetic or nondiabetic origin, when injected into the vitreous of diabetic mice with retinal damage, are incorporated into retinal blood vessels and restore perfusion to ischemic areas. Our studies also show that iPSCs from diabetic donors are able to function in vivo and that reprogramed diabetic iPSC cells behave similarly to nondiabetic hiPSCs. The iPSC-derived ECFCs improved the electroretinograms of the diabetic mice and their ocular kinetic responses. These studies support the notion that iPSCs of diabetic and nondiabetic origin, when differentiated into ECFCs, can correct vascular dysfunction, which in turn improves key functions of the neural retina.

Author Contributions: Conceptualization: C.-H.G., D.C., N.P., M.C.Y., M.E.B. and M.B.G.; methodology: C.-H.G., D.C., N.P., K.B., Y.L., H.-M.C., M.C.Y. and M.B.G.; investigation: C.-H.G., D.C., C.P.V., N.P., S.L.C., S.D.F., P.H., C.H., M.S.S., Y.L., X.H., M.D.D., J.L.F., R.P., A.L.F.L., T.J.M., M.E.B., M.J. and D.N.K.; writing—original draft: C.-H.G., D.C., N.P., M.C.Y. and M.B.G.; writing—review and editing: C.-H.G., D.C., D.N.K., M.E.B., M.C.Y. and M.B.G.; funding acquisition: M.E.B., D.N.K., M.P.M., M.C.Y. and M.B.G.; supervision: M.C.Y. and M.B.G. All authors have read and agreed to the published version of the manuscript.

Funding: This research was funded by National Eye Institute (NEI) [EY012601].

Institutional Review Board Statement: IRB-300000173 University of Alabama.

Informed Consent Statement: Not applicable.

Data Availability Statement: Data will be made available upon request.

Conflicts of Interest: M.C.Y. and N.P. are inventors on a pending patent US20190211304A1 related to this work filed by Indiana University Research and Technology Corp (no. 16/323,722, filed 4 August 2017, published 11 July 2019). M.C.Y. is the scientific founder and holds equity in Vascugen Inc., a biotechnology company deriving regenerative medicine cell therapies from iPSCs.

Reference

1. Gil, C.-H.; Chakraborty, D.; Vieira, C.P.; Prasain, N.; Calzi, S.L.; Fortmann, S.D.; Hu, P.; Banno, K.; Jamal, M.; Huang, C.; et al. Specific mesoderm subset derived from human pluripotent stem cells ameliorates microvascular pathology in type 2 diabetic mice. *Sci Adv.* **2022**, *8*, eabm5559. [CrossRef] [PubMed]

MDPI
St. Alban-Anlage 66
4052 Basel
Switzerland
Tel. +41 61 683 77 34
Fax +41 61 302 89 18
www.mdpi.com

Biology and Life Sciences Forum Editorial Office
E-mail: blsf@mdpi.com
www.mdpi.com/journal/blsf